Prevention through Design

Reinforced Concrete Design

Instructor's Manual

DEPARTMENT OF HEALTH AND HUMAN SERVICES
Centers for Disease Control and Prevention
National Institute for Occupational Safety and Health

Disclaimer

Mention of any company or product does not constitute endorsement by NIOSH. In addition, citations to Web sites external to NIOSH do not constitute NIOSH endorsement of the sponsoring organizations or their programs or products. Furthermore, NIOSH is not responsible for the content of these Web sites.

Ordering Information

This document is in the public domain and may be freely copied or reprinted. To receive NIOSH documents or other information about occupational safety and health topics, contact NIOSH at

Telephone: 1–800–CDC–INFO (1–800–232–4636)
TTY: 1–888–232–6348
Web site: www.cdc.gov/info

or visit the NIOSH Web site at www.cdc.gov/niosh

For a monthly update on news at NIOSH, subscribe to *NIOSH eNews* by visiting www.cdc.gov/niosh/eNews.

DHHS (NIOSH) Publication No. 2013–135

July 2013

SAFER • HEALTHIER • PEOPLE™

Please direct questions about these instructional materials to the National Institute for Occupational Safety and Health (NIOSH):

Telephone: (513) 533–8302
E-mail: preventionthroughdesign@cdc.gov

Foreword

A strategic goal of the Prevention through Design (PtD) Plan for the National Initiative is for designers, engineers, machinery and equipment manufacturers, health and safety (H&S) professionals, business leaders, and workers to understand the PtD concept. Further, they are to apply these skills and this knowledge to the design and redesign of new and existing facilities, processes, equipment, tools, and organization of work. In accordance with the PtD Plan, this module has been developed for use by educators to disseminate the PtD concept and practice within the undergraduate engineering curricula.

John Howard, M.D.
Director, National Institute for
  Occupational Safety and Health
Centers for Disease Control and Prevention

Contents

Acknowledgments

Authors:

John Gambatese, Ph.D., P.E.
Ryan Lujan

The authors thank the following for their reviews:

NIOSH Internal Reviewers

Pamela E. Heckel, Ph.D., P.E.
Donna S. Heidel, M.S., C.I.H.
Thomas J. Lentz, Ph.D., M.P.H.
Rick Niemeier, Ph.D.
Andrea Okun, Ph.D.
Paul Schulte, Ph.D.
Pietra Check, M.P.H.
John A. Decker, Ph.D.
Matt Gillen, M.S., C.I.H.
Roger Rosa, Ph.D.

Peer and Stakeholder Reviewers

Don Bloswick, Ph.D.
COL Daisie D. Boettner, Ph.D.
Joe Fradella, Ph.D.
Matthew Marshall, Ph.D.
Gopal Menon, P.E.
James Platner, Ph.D.
Georgi Popov, Ph.D.
Deborah Young-Corbett, Ph.D., C.I.H., C.S.P., C.H.M.M.

Introduction

This Instructor's Manual is part of a broad-based multi-stakeholder initiative, Prevention through Design (PtD). This module has been developed for use by educators to disseminate the PtD concept and practice within the undergraduate engineering curricula. Prevention through Design anticipates and minimizes occupational safety and health hazards and risks* at the design phase of products,† considering workers through the entire life cycle, from the construction workers to the users, the maintenance staff, and, finally, the demolition team. The engineering profession has long recognized the importance of preventing occupational safety and health problems by designing out hazards. Industry leaders want to reduce costs by preventing negative safety and health consequences of poor designs. Thus, owners, designers, and trade contractors all have an interest in the final design.

This manual is for one of four PtD education modules to increase awareness of construction hazards. The modules support undergraduate courses in civil and construction engineering. The four modules cover the following:

1. **Reinforced concrete design**
2. Mechanical-electrical systems
3. Structural steel design
4. Architectural design and construction.

The manual is specific to a PowerPoint slide deck related to Module 1, **Reinforced concrete design**. It contains learning objectives, slide-by-slide lecture notes, case studies, test questions, and references. It is assumed that the users are experienced professors/lecturers in schools of engineering. As such, the manual does not provide specifics on *how* the materials should be presented. Slide notes are included on most of the slides for the instructor's consideration.

Numerous examples of inadequate design and catastrophic failures can be found on the Internet. If time permits, have the students seek, share, and analyze appropriate and inadequate designs. The PtD Web site is located at www.cdc.gov/niosh/topics/ptd. The National Institute for Occupational Safety and Health (NIOSH) Fatality Assessment and Control Evaluation (FACE) Reports can be found at www.cdc.gov/niosh/face/. Occupational Safety and Health Administration (OSHA) Fatal Facts are available at www.osha.gov/OshDoc/toc_FatalFacts.html.

---

*A "hazard" is anything with the potential to do harm. A "risk" is the likelihood of potential harm from that hazard being realized.
†The term *products* under the Prevention through Design umbrella pertains to structures, work premises, tools, manufacturing plants, equipment, machinery, substances, work methods, and systems of work.

Learning Objectives and Overview

Photo courtesy of Thinkstock

Reinforced Concrete Design
EDUCATION MODULE

Developed by John Gambatese, Ph.D., P.E.
Ryan Lujan
School of Civil and Construction Engineering
Oregon State University

NOTES TO INSTRUCTORS

This module presents safe-design considerations pertaining to reinforced concrete design and construction. It contains specific examples of common workplace hazards related to construction and illustrates ways design can make a difference. A case study is included to facilitate class discussions. One section of slides presents the Prevention through Design (PtD) concept, another summarizes reinforced concrete design principles, and a third illustrates applications of the PtD concept to real-world construction scenarios.

This education module is intended to facilitate incorporation of the PtD concept into your concrete design course. You may wish to supplement the information presented in this module and may assign projects, class presentations, or homework as time permits. Sections may be presented independently of the whole. Presentation times are approximate, based on our presentation experience.

To activate features embedded in some slides, please "enable content," make this a "trusted document," and view the slides in "slide show" mode. To show the presentation file in slideshow mode, press F5. Each slide is accompanied by speaker notes that you can read aloud while the slide is projected on the screen. The audience does not see the speaker notes. When you click on "Use Presenter View" on the Slide Show tab, your monitor displays the speaker notes but the projected image does not.

Thank you for using this module. To report problems or to make suggestions, please contact the National Institute for Occupational Safety and Health (NIOSH):

Telephone: (513) 533–8302
E-mail: preventionthroughdesign@cdc.gov

SOURCE
Photo courtesy of Thinkstock

Guide for Instructors

Topic	Slide numbers	Approx. minutes
Introduction to Prevention through Design (PtD)	5–29	45
Elements, Activities, and Hazards	30–45	30
Mitigating Concrete Construction Hazards	46–78	50
Construction Case Study	79–82	20
Recap	83–84	5
References and Other Sources	85–95	—

Reinforced Concrete

NOTES

The first two slides of the presentation provide acknowledgments and general information. Learning objectives are delineated on Slide 3. Slide 4 contains the Overview. Slides 5 through 29 introduce the PtD concept and can be covered in approximately 45 minutes. Slides 30 through 45 use a variety of concrete design references to summarize the concrete design and detailing process, as well as concrete construction activities. Some instructors will wish to use the two sets of process slides earlier in their course, well before they explicitly cover PtD. Construction safety and hazards are rarely covered in engineering curricula but are important for engineers to understand if they are to implement PtD. Slides 46 through 78 are provided to educate students about the hazards associated with concrete construction. In addition, these slides provide specific examples of PtD opportunities in reinforced concrete designs. Finally, a construction case study is provided in slides 79 through 82 to illustrate how design impacts construction safety and how PtD could be implemented to prevent injuries and fatalities.

Learning Objectives

- Explain the Prevention through Design (PtD) concept.

- List reasons why project owners may wish to incorporate PtD in their projects.

- Identify workplace hazards and risks associated with design decisions and recommend design alternatives to alleviate or lessen those risks.

Reinforced Concrete

NOTES

At the completion of this education module, an engineering student should be able to

- Explain the PtD concept.
- Describe motivations, barriers, and enablers for implementing PtD on projects.
- List three reasons why PtD improves business value.

Overview

- PtD Concept

- Introduction to Reinforced Concrete

- Reinforced Concrete Design Process, Construction Activities, and Safety Hazards

- Reinforced Concrete PtD Examples

- Case Study

Photo courtesy of Thinkstock

Reinforced Concrete

NOTES

This is an overview of the PtD topics that we will cover in this module. Many of you are probably not familiar with PtD. We will discuss the concept. Next I will summarize the concrete design, detailing, and manufacturing processes. Similarly, I will summarize the concrete construction process, which has also come up in class previously. Finally, I will show you some specific ways that design engineers can incorporate PtD into their concrete designs. A case study will be used to illustrate how PtD concepts can be used to improve safety in reinforced concrete construction.

SOURCE

Photo courtesy of Thinkstock

Introduction to
Prevention through Design (PtD)

Introduction to Prevention through Design

EDUCATION MODULE

NOTES

Let's start by introducing PtD.

Occupational Safety and Health

- Occupational Safety and Health Administration (OSHA) www.osha.gov
  - Part of the Department of Labor
  - Assures safe and healthful workplaces
  - Sets and enforces standards
  - Provides training, outreach, education, and assistance
  - State regulations possibly more stringent

- National Institute for Occupational Safety and Health (NIOSH) www.cdc.gov/niosh
  - Part of the Department of Health and Human Services, Centers for Disease Control and Prevention
  - Conducts research and makes recommendations for the prevention of work-related injury and illness

Reinforced Concrete

NOTES

All employers, including structural design firms, are required by law to provide their employees with a safe work environment and training to recognize hazards that may be present. They also must provide equipment or other means to minimize or manage the hazards.

Designers historically have not been familiar with the federal Occupational Safety and Health Act (OSH Act) standards because they were rarely exposed to construction jobsite hazards. However, with the increasing roles that designers are playing on worksites, such as being part of a design-build team, it is becoming increasingly important that they receive construction safety training, including information about federal and state construction safety standards.

The Occupational Safety & Health Act of 1970, Public Law 91-596 (OSH Act) [29 USC* 1900], was passed on December 29, 1970, "To assure safe and healthful working conditions for working men and women; by authorizing enforcement of the standards developed under the Act; by assisting and encouraging the States in their efforts to assure safe and healthful working conditions; by providing for research, information, education, and training in the field of occupational safety and health; and for other purposes." The construction industry standards

---

*United States Code. See USC in Sources.

enforced by the Occupational Safety and Health Administration (OSHA) are found in Title 29 Part 1926 of the Code of Federal Regulations [29 CFR 1926].

The National Institute for Occupational Safety and Health (NIOSH) is part of the Department of Health and Human Services, Centers for Disease Control and Prevention. The National Occupational Research Agenda (NORA) is a partnership program to stimulate innovative research and improved workplace practices. Unveiled in 1996, NORA has become a research framework for NIOSH and the nation. Diverse parties collaborate to identify the most critical issues in workplace safety and health. Partners, then, work together to develop goals and objectives for addressing these needs. Participation in NORA is broad, including stakeholders from universities, large and small businesses, professional societies, government agencies, and worker organizations. NIOSH and its partners have formed ten NORA Sector Councils: Agriculture, Forestry & Fishing; Construction; Healthcare & Social Assistance; Manufacturing; Mining; Oil and Gas Extraction; Public Safety; Other Services; Transportation, Warehousing & Utilities; and Wholesale and Retail Trade. The mission of the NIOSH research program for the Construction sector is to eliminate occupational diseases, injuries, and fatalities among individuals working in these industries through a focused program of research and prevention.

SOURCES

CFR. Code of Federal Regulations. Washington, DC: U.S. Government Printing Office, Office of the Federal Register.

NIOSH FACE reports [www.cdc.gov/niosh/face]

OSHA Fatal facts accident reports index [www.setonresourcecenter.com/MSDS_Hazcom/FatalFacts/index.htm]

OSHA home page [www.osha.gov]

USC. United States Code. Washington, DC: U.S. Government Printing Office.

Construction Hazards

 **Construction Hazards**

- Cuts
- Electrocution
- Falls
- Falling objects
- Heat/cold stress
- Musculoskeletal disease
- Tripping

[BLS 2006; Lipscomb et al. 2006]

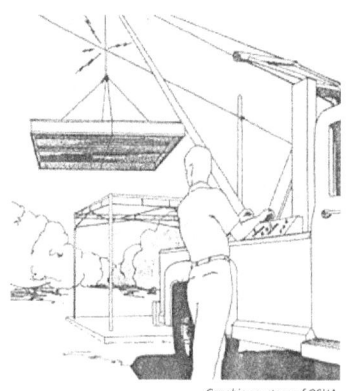

Graphic courtesy of OSHA

CDC *NIOSH*

Reinforced Concrete

NOTES

A construction worksite by its nature involves numerous potential hazards. A portion of the work is directly affected by weather. Workers interact with heavy equipment and materials at elevated heights, in below-ground excavations, and in multiple awkward positions. The composition of the site workforce changes over the project, and work is done autonomously at times and in coordination at others. The construction worksite is unforgiving to poor planning and operational errors.

For these reasons, pre-job construction-phase planning is used as a best practice to systematically address potential hazards. Project-specific worker safety orientations prior to site work also play an important role. PtD practices, by systematically looking further upstream at design-related potential hazards, extend these pre-job measures. PtD can help identify potential hazards so that they can be eliminated, reduced, or communicated to contractors for pre-job planning.

Every hazard that can be addressed should be addressed. Falling can cause serious injury. Boilermakers, pipe-fitters, and iron workers can experience career-ending musculoskeletal injuries by lifting heavy loads or working in a cramped position. Anyone can be seriously injured by a falling object. Whether a structural member or a simple wrench, a falling object can be deadly. Anyone can trip, but the elevated height and proximity to dangerous equipment increase the risk of injury on a construction site. Some accidents are caused by poor lighting and/or sunlight glare. Common injuries due to spatial misperception include hitting your head or cutting yourself on sharp corners. Hot summer and cold winter days can affect worker health. Personal protective equipment (PPE), such as hardhats, gloves, ear protection, and safety glasses, is required for a reason! Not every hazard on a construction worksite can be "designed out," but many significant ones can be minimized during the design phase.

SOURCES

BLS [2006]. Injuries, illnesses, and fatalities in construction, 2004. By Meyer SW, Pegula SM. Washington, DC: U.S. Department of Labor, Bureau of Labor Statistics, Office of Safety, Health, and Working Conditions.

Lipscomb HJ, Glazner JE, Bondy J, Guarini K, Lezotte D [2006]. Injuries from slips and trips in construction. Appl Ergonomics 37(3):267–274.

OSHA [ND]. Fatal facts accident reports index [foreman electrocuted]. Accident summary no. 17 [www.setonresourcecenter.com/MSDS_Hazcom/FatalFacts/Index.htm].

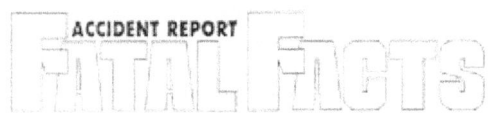

ACCIDENT SUMMARY No. 17

Accident Type:	Electrocution
Weather Conditions:	Sunny, Clear
Type of Operation:	Steel Erection
Size of Work Crew:	3
Collective Bargaining	No
Competent Safety Monitor on Site:	Yes - Victim
Safety and Health Program in Effect:	No
Was the Worksite Inspected Regularly:	Yes
Training and Education Provided:	No
Employee Job Title:	Steel Erector Foreman
Age & Sex:	43-Male
Experience at this Type of Work:	4 months
Time on Project:	4 Hours

BRIEF DESCRIPTION OF ACCIDENT

Employees were moving a steel canopy structure using a "boom crane" truck. The boom cable made contact with a 7200 volt electrical power distribution line electrocuting the operator of the crane; he was the foreman at the site.

INSPECTION RESULTS

As a result of its investigation. OSHA issued citations for four serious violations of its construction standards dealing with training, protective equipment, and working too close to power lines.

OSHA's construction safety standards include several requirements which, If they had been followed here. might have prevented this fatality.

ACCIDENT PREVENTION RECOMMENDATIONS

1. Develop and maintain a safety and health program to provide guidance for safe operations (29 CFR 1926.20(b)(1)).
2. Instruct each employee on how to recognize and avoid unsafe conditions which apply to the work and work areas (29 CFR 1926.21(b)(2))
3. If high voltage lines are not de-energized, visibly grounded, or protected by insulating barriers, equipment operators must maintain a minimum distance of 10 feet between their equipment and the electrical distribution or transmission lines (29 CFR 1926.550(a)(15)(i)).

SOURCES OF HELP

- Ground Fault Protection on Construction Sites (OSHA 3007) which describes OSHA requirements for electrical safety at construction sites.

- Construction Safety and Health Standards (OSHA 2207) which contains all OSHA job safety and health rules and regulations (1926 and 1910) covering construction
- OSHA Safety and Health Training Guidelines for Construction (available from the National Technical Information Service - Order No PB-239312/AS) comprised of a set of 15 guidelines to help construction employees establish a training program in the safe use of equipment, tools, and machinery on the job
- OSHA-funded free onsite consultation services Consult your telephone directory for the number of your local OSHA area or regional office for further assistance and advice (listed under the US Labor Department or under the state government section where states administer their own OSH programs).

NOTE: The case here described was selected as being representative of fatalities caused by improper work practices. No special emphasis or priority is implied nor is the case necessarily a recent occurrence. The legal aspects of the incident have been resolved, and the case is now closed.

Construction Accidents

Construction Accidents in the United States

Construction is one of the most hazardous occupations. This industry accounts for

- 8% of the U.S. workforce, but 20% of fatalities

- About 1,100 deaths annually

- About 170,000 serious injuries annually

[CPWR 2008]

Photo courtesy of Thinkstock

Reinforced Concrete

NOTES

As many of us know, construction is one of the most dangerous industries for workers. In the United States, construction typically accounts for 170,000 serious injuries and 1,100 deaths each year. The fatality rate is disproportionally high for the size of the construction workforce. Twenty percent of all collapses during construction are the result of structural design errors. Statistics like these do not tell the whole story. Behind every serious injury, there is a real story of an individual who suffered serious pain and may never fully recover. Behind every fatality, there are spouses, children, and parents who grieve every day for their loss. We all recognize that safety is a vital component of an inherently dangerous business. All of us—including architects and engineers—must do what we can to reduce the risk of injuries on our projects.

SOURCES

CPWR [2008]. The construction chart book. 4th ed. Silver Spring, MD: Center for Construction Research and Training.

New York State Department of Health [2007]. A plumber dies after the collapse of a trench wall. Case report 07NY033 [www.cdc.gov/niosh/face/pdfs/07NY033.pdf].

OSHA [ND]. Fatal facts accident reports index [laborer struck by falling wall]. Accident summary no. 59 [www.setonresourcecenter.com/MSDS_Hazcom/FatalFacts/index.htm].

Photo courtesy of Thinkstock

NEW YORK
state department of
HEALTH

FATALITY ASSESSMENT AND CONTROL EVALUATION

A Plumber Dies After the Collapse of a Trench Wall
Case Report: 07NY033

SUMMARY

In May 2007, a 46 year old self-employed plumbing contractor (the victim) died when the unprotected trench he was working in collapsed. The victim was an independent plumber subcontracted to install a sewer line connection to the sewer main, part of a general contractor project to install a new sanitary sewer for an existing single family residence.

At approximately 12:30 PM on the day of the incident, the workers on site observed the victim walking back toward the residence for parts as they initiated their lunch break. When the victim did not come for his lunch or answer his cell phone, the general contractor and workers starting searching for the victim. The excavation contractor observed that a portion of the trench had collapsed where the victim was installing a sewer tap. The victim was found trapped in the trench under a large slab of asphalt, rock and soil. Three workers immediately climbed down the side of the trench to try to assist the victim. One of the workers called 911 on his cell phone. Police and emergency medical services (EMS) arrived on site within minutes. The EMS members entered the unprotected trench but could not revive the victim. The county trench rescue team recovered the victim's body at approximately seven feet below grade and lifted him from the ditch four hours after the incident. He was pronounced dead at the site. More than 50 rescue workers were involved in the recovery.

New York State Fatality Assessment and Control Evaluation (NY FACE) investigators concluded that, to help prevent similar occurrences, employers and independent contractors should:
- **Require that all employees, subcontractors, and site workers working in trenches five feet or more in depth are protected from cave-ins by an adequate protection system.**
- **Require that a competent person conducts daily inspections of the excavations, adjacent areas, and protective systems and takes appropriate measures necessary to protect workers.**
- **Require that all employees and subcontractors have been properly trained in the recognition of the hazards associated with excavation and trenching. In addition, the general contractor (GC) should be responsible for the collection and review of training records and require that all workers employed on the site have received the requisite training to meet all applicable standards and regulations for the scope of work being performed.**
- **Require that on a multi-employer work site, the GC should be responsible for the coordination of all high hazard work activities such as excavation and trenching.**

- **Require that all employees are protected from exposure to electrical hazards in a trench.**

Additionally,

- **Employers of law enforcement and EMS personnel should develop trench rescue procedures and should require that their employees are trained to understand that they are not to enter an unprotected trench during an emergency rescue operation.**
- **Local governing bodies and codes enforcement officers should receive additional training to upgrade their knowledge and awareness of high hazard work, including excavation and trenching. This skills upgrade should be provided to both new and existing codes enforcement officers.**
- **Local governing bodies and codes enforcement officers should consider requiring building permit applicants to certify that they will follow written excavation and trenching plans in accordance with applicable standards and regulations, for any projects involving excavation and trenching work, before the building permits can be approved.**

INTRODUCTION

In May, 2007, a 46 year old self-employed plumbing contractor died when the trench he was working in collapsed at a residential construction site. Approximately 8000 pounds of broken asphalt, rock and dirt fell atop the victim, fatally crushing him as he was installing a sewer tap to a town sewer main. The New York State Fatality Assessment and Control Evaluation (NY FACE) program learned about the incident from a newspaper article the following day. The Occupational Safety and Health Administration (OSHA) investigated the incident along with the county sheriff's office. The NY FACE staff met and reviewed the case information with the OSHA compliance officer. This report was developed based upon the information provided by OSHA, the county sheriff's department, and the county coroner's medical and toxicological reports.

The general contractor (GC) on the residential construction site had been hired by the homeowners to complete a project that included the installation of a new sanitary sewer connection for an existing single family residence. The GC was the owner and sole employee of his company, which had been in business for many years. The GC directed the work of two subcontractors on the work site to complete the installation of the residential sewer line.

- One subcontractor was an excavating company that had been in business for approximately four years. The owner of this company hired two workers to assist him with the excavation of the trench.
- The second subcontractor was the victim, a self-employed licensed plumber who had over twenty years of experience with a variety of construction projects, including the installation of sewer lines. The victim did not have any previous work relationship with either the GC or the excavation subcontractor.

The OSHA investigation report indicated that the GC and the subcontractor did not have health and safety programs. A formal health and safety plan had not been established to identify the hazards of the excavation project and the actions to be taken to remediate them. The GC, subcontractors and the subcontractors' employees did not have hazard recognition training or safety training on the fundamentals of excavation and trenching. None of the workers on the site were knowledgeable on excavation and trenching safety standards and applicable regulations and they did not understand the hazards and dangers associated with working in a trench. A competent person was not present to conduct initial and ongoing inspections of the excavation project, identify potential health and safety hazards such as possible cave-in, and oversee the use of adequate protection systems and work practices.

INVESTIGATION

The GC was hired to replace a crushed sewer line that ran under the driveway of an existing single family residence. Rather than dig up the driveway to replace the old line, which was thought to be more costly and time-consuming, the GC decided to run a new line. All required town permits had been obtained and the local codes enforcement requirements for one-call system notification of the excavation and underground utility location mark-outs had been completed. The work had been scheduled to be completed in one day (Friday), but the excavation subcontractor lost time due to hitting a water line and encountering very rocky soil during the excavation. The project had to be extended to two days (Friday and Monday). The town water and sewer inspector visited the work site on Friday, observed the digging of the trench which began at the residence, and halted the digging of the trench at the edge of the property to avoid having an open trench in the road and consequent road closure over a weekend. Excavation company workers had been observed in the trench spotting and hand digging.

On Monday, the day of the incident, the excavating subcontractor initiated excavation from the edge of the road to the sewer main in the roadway. An employee witness of the excavating company stated that the victim was directing excavation work while in the trench and hand digging to expose the sewer main once the excavator came close to the location. OSHA findings indicated that tools were uncovered in the trench in the area of the trench wall collapse, including a shovel, pick ax, hammer drill and drill bits, consistent with the scenario of the victim being in the ditch, hand digging to locate the sewer main. The town water and sewer inspector also visited the work site on Monday. He determined that the victim did not have the correct parts to complete the sewer connection, advised him of the correct parts, and indicated that he would return later in the day to re-inspect and photograph the completed sewer tap in order to allow the excavating subcontractor to run the pipe back to the house, backfill the excavation and reopen the road.

The GC left the work site to purchase the correct parts, while the excavation continued. The dimensions of the final trench were approximately 55 feet in length, 3 feet to 8 feet in depth, and 30 inches to 128 inches in width (see Figure 1). It was shaped like a "T." The gravity sewer main that the victim was connecting to was located at a depth of 7 feet 4 inches (7' 4") below grade at the east (E) end of the top of the "T." Installation of new sewer pipe from the residence had been initiated and some of the trench had already been backfilled. The length of the trench from the top of the "T" to the location of the newly installed sewer pipe was 35 feet 11 inches (35'11") at the time of the incident. Soil analysis results, conducted after the incident, indicated a granular, sandy gravel Type C soil (OSHA Excavation Standard) that contained large cobbles and boulders, the least stable soil type.

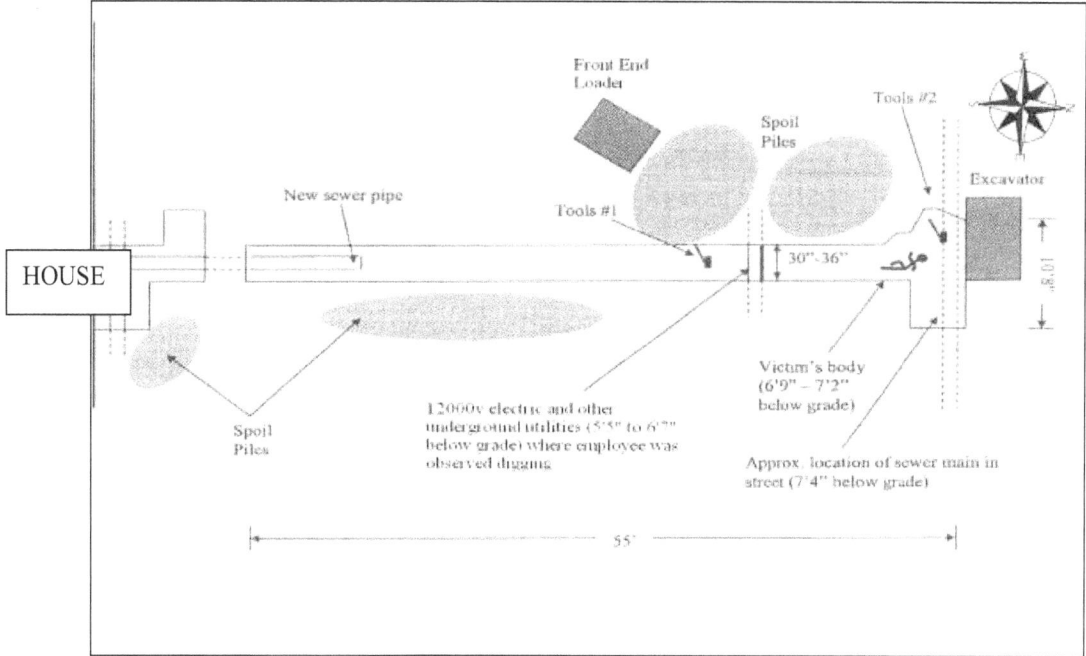

Figure 1: Schematic of the excavation and the incident site (courtesy of OSHA)

The faces of the trench were vertical. No shoring or benching was used. Large cobbles and boulders and loose rock/dirt were visible on the face of the excavation and were not removed or supported. The pavement above the E and W faces of the excavation had been undermined during excavation activities and no support system was utilized to protect employees from a possible collapse. Pieces of road pavement and asphalt had been undermined during excavation activities in the road in the proximity of the sewer main at the top of the "T." These areas were in plain view and did not have additional support. On the W side of the excavation, loose boulders, rock and debris in spoils piles were located less than two feet from the edge of the trench. (Figure 2) The excavator was positioned adjacent to the N end of the trench, where undermined areas were in plain sight. The N end of the trench, where the victim was installing the sewer tap, also lacked an access ladder or other safe means of entry/egress.

Figure 2: View of the west wall of the excavation south of the "T."
Note the boulders and loose rock/dirt on the excavation face as well as the location of the spoils pile within 2 feet of the edge of the trench. (courtesy of OSHA)

The GC returned just before 12 noon with the correct parts and handed them to the victim. The GC left the site in order to purchase lunch for the workers, including the victim. At this same time, the victim called the town water and sewer inspector, informed him that he had located the sewer main, had all the correct parts, and was ready to connect. The town inspector informed the victim that someone from the town would be out after lunch to inspect and photograph the sewer tap. According to the town inspector, a sewer tap to a sewer main is a simple job that would take about 20 minutes to complete. The GC returned with lunch at 12:30PM. The workers, with the exception of the victim, took a break for lunch at a location near the front end loader (Figure 1). The workers saw the victim walking in the trench in the direction of the residence and heard him say that he was "looking for a splitter for a three-way." By 1:00 PM the victim still had not come for his lunch. The GC called the victim on his cell phone and looked for him in his van behind the house. The other workers joined in the search. The excavating subcontractor observed that a portion of the west side of the trench had collapsed. When the workers approached the excavation, they found the victim trapped in the trench under a large slab of asphalt, rock, and soil, with only the back of his head exposed. Three workers climbed down the side of the trench to try to assist the victim.

The workers removed the dirt from around his head, lifted his head, and tried to clear his airway. They checked for a pulse, but found none. One of the workers then called 911 from his cell phone. The workers attempted to move the slab of asphalt without success. Within minutes, the police arrived, followed by EMS at approximately 1:08 PM. The EMS personnel entered the unprotected trench but were unable to revive the victim. Volunteer firefighters from multiple fire departments and a special trench rescue team responded, the latter team having been created by the county after the deaths of two workers in a construction trench collapse 10 years earlier. A wooden safety box was built by the trench rescue team and efforts began to free the victim from entrapment by chipping the asphalt slab into pieces. Using a system of ropes and pulleys, the rescue team lifted the victim from the ditch at 4:25 PM. His body had been recovered at about 7' below grade. The county coroner pronounced him dead at 4:35 PM. Approximately 50 rescuers responded to the 911 call.

The OSHA investigation resulted in findings that the trench section that collapsed was a triangular shaped area at the northwest corner of the excavation, approximately 5 feet 1 inch (5' 1") in length, 4 feet (4') wide, and 6-7 feet (6-7') deep. Multiple hazards were present, but had not been identified and remediated. The W side of the excavation collapsed and pieces of asphalt paving and rock fatally crushed the victim while he was making the sewer tap (Figures 3 and 4).

The hazards of the unprotected trench exposed additional people to the excavation collapse as the GC, the excavation company workers and EMS personnel entered the trench to attempt a rescue of the victim. In addition to the trench hazards, no precautions had been taken to prevent exposure to the underground electrical and utility lines. The town inspector had noted that a young employee of the excavation company was "manually hand digging with shovel and pick ax "within a few inches of the buried electrical lines." This is consistent with OSHA findings that indicated attempts had been made to cut the rock in the face of the trench at the location of the underground utilities. A demo saw, hammer drill and cordless reciprocating saw used to cut rocks and pavement were found within inches of the 12,000 volt underground electrical line. Several other utilities were also exposed in this location at the edge of the road (Figure #1, Tools #1). EMS personnel also entered the trench when power was still connected to the utilities in the trench.

Figure 3: Location of collapse.
Note spoils piles and equipment located less
2 feet from the edge of the trench
(courtesy of OSHA)

Figure 4: Area of trench collapse
Note the large boulders hanging from the than
excavation faces and undermined areas on the
edge of the trench (courtesy of OSHA)

RECOMMENDATIONS/DISCUSSION

Recommendation #1: *Employers and independent contractors should require that all employees, subcontractors and site workers working in trenches five feet or more in depth are protected from cave-ins by an adequate protection system.*

Discussion: Employers and contractors should require that all employees working in trenches five feet deep or more are protected from cave-ins by an adequate protection system appropriate to the conditions of the trench, including sloping techniques or support systems such as shoring or trench boxes (OSHA 29CFR 1926.652). Sloping involves positioning the soil away from an excavation trench at an angle that would prevent the soil from caving into the trench. Even in shallow trenches less than five feet in depth, the possibility of accidents still exists. Trenches five feet deep or less should also be protected if a competent person identifies a cave-in potential. Trench protection systems are available to all employers and independent contractors, even as rental equipment. Employers should also require that all pieces of excavated pavement, asphalt, dirt, rock, boulders, and debris as well as excavation equipment are located in spoils piles or positions that are at least two feet from the edge of the excavated trench. Where a two foot setback is not possible, spoils may need to be hauled to another location. In this incident, sloping would not have been an appropriate protection system, due to the composition of the soil. Employers and contractors should consult tables located in the appendices of the OSHA Excavation Standard that detail the protection required based upon the soil type and environmental conditions present at a work site. Employers and contractors can also consult with manufacturers of protective systems to obtain detailed guidance for the appropriate use of protection systems.

Trenches should be kept open only for the minimum amount of time needed. Hinze and Bren (1997) observed that the risk of a collapse in an unprotected trench increases the longer a trench is open. They propose that after a trench is dug, the apparent cohesion of trench walls may begin to relax after only four hours, contributing to increasingly unstable walls in an unprotected trench. In this incident, a 45 feet length of the trench had been excavated and was left open for more than two days. The trench section where the incident occurred was dug at approximately 8:30 AM on the day of the incident. Hand digging and incorrect parts resulted in additional delays in making the sewer tap to the main. The trench collapse occurred approximately four hours later, between 12:30 PM and 1:00 PM.

The key to preventing a trench accident is not to enter an unprotected trench. When the walls of a trench collapse or cave in, the results are entrapment or struck-by incidents to anyone caught inside, accidents which can occur in seconds. Many workers in a trench are in a kneeling or squatting position that results in little opportunity for an escape. Victims do not need to be completely covered in soil. Even with partial covering, enough pressure is created for mechanical asphyxia in which the weight of the dirt and soil compresses the chest. One cubic yard of soil has an average weight of 2500 pounds (Figure 4), but can vary due to the composition and moisture content.

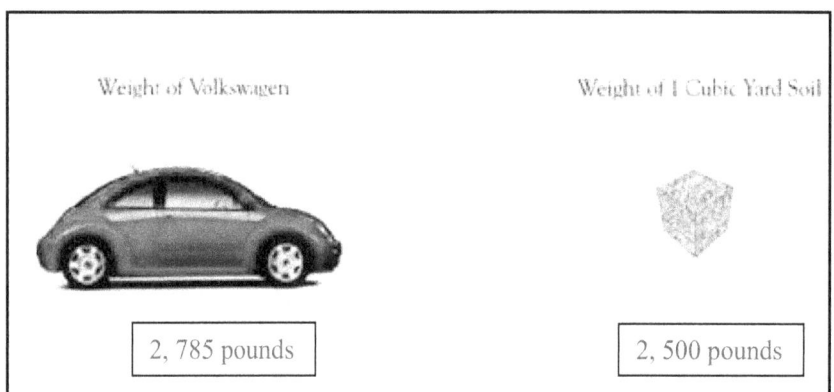

Figure 5: Weight of one cubic yard of soil (courtesy of "Weights of Building Materials, Agricultural Commodities, and Floor Loads for Buildings" standard reference)

Recommendation #2: *Employers and independent contractors should require that a competent person conducts daily inspections of the excavations, adjacent areas, and protective systems and takes appropriate measures necessary to protect workers.*

Discussion: Employers and independent contractors are responsible for complying with the OSHA Excavation Standard requirements to designate a competent person on site for excavation and trenching projects to make daily inspections of excavations, the adjacent areas, and protective systems (OSHA 29CFR 1926.651). A competent person is defined as someone who is capable of identifying existing and predictable hazards in the surroundings and working conditions that are dangerous to employees and who has the authorization to take prompt corrective measures to eliminate them. They should inspect the trenches daily, as needed throughout the work shift, and as conditions change (for example, heavy rainfall or increased traffic vibrations). These inspections should be conducted before worker entry, to ensure that there is no evidence of a possible cave-in, failure of a protective system, hazardous conditions such as spoils piles or equipment location, or hazardous atmosphere.

In particular, competent persons are required by OSHA to complete a competent person training curriculum, which could be an OSHA training program or an equivalent safety or trade organization training. The competent person needs be knowledgeable on the hazards associated with excavation and trenching, as well as the causes of injuries and the safe work practices and specific protective actions needed. Competent persons must also be experienced in excavation and trenching with a minimum of hands-on training in a demonstration trench or in a field component. The competent person needs to know the key points of the OSHA Excavation Standard, including the excavation standards and appendices, checklists, soils analysis and the components of a daily trenching inspection.

Having a competent person is a particularly acute problem among contracting companies that employ fewer than 10 workers. Of the National Institute for Occupational Safety and Health (NIOSH) FACE cases related to excavation and trenching, 88% were non-union companies with less than 10 workers. These small companies are not members of trade associations and are the least likely to employ trench safety protections and to have an adequately trained competent person or an excavation crew.

In this incident, no competent person was hired by the GC to conduct initial and ongoing inspections of the trench. The GC, excavating contractor, and excavation company employees did not possess an understanding of the hazards associated with excavation and trenching operations or a knowledge of the requirements of the OSHA Excavation Standard. No one on-site was qualified to function as the competent person.

Recommendation #3: *Employers and independent contractors should require that all employees and subcontractors have been properly trained in the recognition of the hazards associated with excavation and trenching. On a multi-employer work site, the GC should be responsible for the collection and review of training records and require that all workers employed on the site have received the requisite training to meet all applicable standards and regulations for the scope of work being performed.*

Discussion: Excavation and trenching is one of the most hazardous construction operations. Even with a competent person on site, workers in excavation and trenching operations are also in need of health and safety training, including basic hazard recognition and prevention. Workers should be able to identify the specific hazards associated with excavation and trenching, the reasons for using protective equipment and how to work in a trench safely. Workers should be trained not to enter an unprotected trench, even in a rescue attempt, since they place themselves at risk of becoming injured or killed. If necessary, projects should be delayed until training requirements are met and training records are provided.

In this case, the general contractor, excavation subcontractor, and excavation company employees did not demonstrate adequate knowledge of safe work practices in excavation and trenching. The limited training in proper excavation technique as well as inadequate hazard recognition and prevention training were critical to the failure to properly assess the hazards present and protect the trench.

Recommendation #4: *Employers and independent contractors should require that on a multi-employer work site, the GC should be responsible for the coordination of all high hazard work activities such as excavation and trenching.*

Discussion: The GC is responsible and accountable for the safety of all employees, subcontractors and workers on the site. Health and safety plans should be in place to formally address the hazards that may be encountered, including written plans to manage these hazards and protect the safety of all workers on the site.

In this incident, the GC did coordinate the work activities of the subcontractors and workers on the job, but health and safety plans were not addressed. The management of excavation and trenching hazards was left to a subcontractor who was not a competent person, knowledgeable or trained in the requirements of the OSHA Excavation Standard.

Recommendation #5: *Employers of law enforcement and EMS personnel should develop trench rescue procedures and should require that their employees are trained to understand that they are not to enter an unprotected trench during an emergency rescue operation.*

Discussion: Employers of law enforcement and EMS personnel should develop a formal safety procedure for emergency rescue in an unprotected trench. Entering an unprotected trench after a cave-in or collapse could place would-be rescuers in danger. Rescue is a delicate and slow operation requiring knowledge of the behavior of unstable soil, necessary to prevent further injury to the victim or the rescuers. The added weight and vibrations can also contribute to an increased susceptibility to further collapse. Many rescuers precipitate second and third stage trench cave-ins and have become victims themselves. In this incident EMS personnel entered the unprotected trench in an attempt to rescue the victim, exposing themselves to an excavation collapse hazard.

Emergency rescue workers, such as law enforcement officials and EMS personnel, should receive specialized training in how to rescue workers who may be trapped in utility trenches, and should not put themselves in danger by entering an unprotected trench. In this incident, a specialized rescue team was called in to respond to the emergency. The rescue workers had special equipment for trench rescues and building collapses and had undergone specialized training in the area of trench/building collapse emergencies. They immediately constructed a wooden safety box in the trench with a system of ropes and pulleys before entering the trench to free the victim. National Fire Protection Association (NFPA) 1670, Chapter 11 details the requirements for rescue operations after a trench cave-in occurs.

Recommendation #6: *Local governing bodies and codes enforcement officers should receive additional training to upgrade their knowledge and awareness of high hazard work, including excavation and trenching. This skills upgrade should be provided to both new and existing codes enforcement officers.*

Discussion: This recommendation may create a mechanism of observation and oversight by the codes enforcement officers who are likely to encounter small employers and independent contractors during their work. The officers could inform the employers and contractors of potential hazards, provide fact sheets that highlight the key requirements for the excavation and trenching standards, and check some of the basics of the trenching project such as depth of the trench, protection of the trench and identification of the competent person. In addition, they could advise employers and contractors to contact safety experts to learn about and implement trench safety. This may be an effective accident prevention strategy, reaching the thousands of untrained and unprepared small employers and independent contractors with awareness and guidance, the very workers who represent the major group of fatalities in New York State.

In this incident, the town water and sewer inspector observed workers in the unprotected trench serving as spotters, observed a worker hand digging within a few feet of a live buried electrical utility, and observed the victim spotting in the unprotected trench for the excavating subcontractor while attempting to locate the sewer main. If the above recommendation was in place, with a trained and knowledgeable officer, at a minimum the excavation work may have been halted and entry into an unprotected trench may have been prohibited.

Recommendation #7: *Local governing bodies and codes enforcement officers should consider requiring building permit applicants to certify that they will follow written excavation and trenching plans in accordance with applicable standards and regulations, for any projects involving excavation and trenching work, before the building permits can be approved.*

Discussion: Local governing bodies may consider revising building permits to require building permit applicants to certify that they will follow written plans for any projects involving excavation and trenching. Statements on the permit applications would be added to indicate that the employer/independent contractor agrees to accept and abide by all standards and regulations governing the excavation and trenching work, not just local governing body codes and ordinances. If construction companies and independent contractors were required to provide written documentation of how the high hazard work of excavation and trenching will be performed safely as part of the building permit application process, it may prompt the employers and contractors to plan ahead, formally assess the hazards, seek assistance in developing the required safety and injury prevention program, and implement the necessary injury prevention measures. No work should be initiated unless these requirements are met after review and approval. These changes may help to prevent trench related fatalities in NYS.

Recommendation #8: *Employers and independent contractors should require that all employees are protected from exposure to electrical hazards in a trench.*

Discussion: Utilities to the single family residence were located underground in the trench near the edge of the road. Workers were observed using power and hand tools within inches of live 12,000 volt lines. This did not contribute to the fatality, but did present another potential hazard to workers in the excavation and trenching project and to the rescue workers. Performing cutting work next to hot utility lines could have resulted in additional serious injuries and death from electrocution. The company performed the utility mark-out as required by local codes but did not contact the utility company to turn off the power as required, when they realized the need to hand cut large rocks and boulders in the trench. The power was not shut off to these lines until after the incident, when workers returned to complete the work.

Key words: Trench, collapse, cave-in, trenching, excavation, trench protection systems, entrapment, spoils piles

REFERENCES:

1. Associated General Contractors of America Safety Training for the Focus Four. *Hazards in Construction.* Retrieved February 8, 2011 from http://www.agc.org/cs/career_development/safety_training/focus_four_locations

2. CDC/NIOSH. *NIOSH Safety and Health Topic: Trenching and Excavation.* Retrieved on February 8, 2011 from http://www.cdc.gov/niosh/topics/trenching/

3. CDC/NIOSH. MMWR. 2004. *Occupational Fatalities During Trenching and Excavation Work - United States, 1992-2001*. Morbidity and Mortality Weekly Report, 53(15):311-314. Retrieved February 8, 2011 from www.cdc.gov/mmwr/preview/mmwrhtml/mm5315a2.htm

4. CDC/NIOSH. Alert: July 1985. *Preventing Deaths and Injuries from Excavation Cave-ins.* retrieved February 8, 2011 from http://www.cdc.gov/niosh/docs/85-110

5. CDC/NIOSH. Fatality Assessment and Control Evaluation (FACE) investigation reports. Retrieved February 8, 2011 from www.cdc.gov/niosh/face

6. Center to Protect Workers' Rights (CPWR). Plog, Barbara et al. March, 2006. *Barriers to Trench Safety: Strategies to Prevent Trenching-Related Injuries and Deaths.* Retrieved February 8, 2011 from www.elcosh.org.

7. Commonwealth of Massachusetts. Executive Office of Labor and Workforce Development. *Trenching Hazard Alert for Public Works Employers and Employees in Massachusetts.* Bulletin 407, 11/2007, p1-4.

8. Deatherage, J.H., et al. 2004 *Neglecting Safety Precautions may lead to trenching fatalities*. American Journal of Industrial Medicine, 45(6):522-7.

9. EC&M online. June, 2009. *Danger Uncovered.* Beck, Ireland. Retrieved February 8, 2011 from http://ecmweb.com/construction/electrical-trench-safety-20090601/

10. Encyclopedia of Occupational Health and Safety. 4th Edition. *Chapter 93: Construction Trenching* by Jack Mickle. *Types of Projects and Their Associated Hazards* by Jeffrey Hinkman. Retrieved February 8, 2011 from
    http://www.elcosh.org/en/document/296/d000279/encyclopedia-of-occupational-safety-%2526-health-%253A-chapter-93-construction.html

11. Executive Safety Update. The Monthly News Bulletin of the Construction Safety Center, Vol. 17, Issue 3, September, 2009

12. Hinze, J.W. and K. Bren. 1997. *The causes of trenching-related fatalities*. Construction Congress V: Managing Engineered Construction in Expanding Global Markets. Proceedings of the Congress, sponsored by the American Society of Civil Engineers (ASCE), 131(4): 494-500.

13. Irizarry, J. et al: 2002 *Analysis of Safety Issues in Trenching Operation.* 10th Annual Symposium on Construction Innovation and Global Competitiveness, September 9-13, 2002. Retrieved February 8, 2011 from Construction Safety Alliance site: http://engineering.purdue.edu/CSA/publications/trenching03

14. Job Health and Safety Quarterly. Fall, 2009. *Trenching is a Dangerous and Dirty Business.* Retrieved February 8, 2011 from http://www.elcosh.org/en/document/161/d000168/trenching-is-a-dangerous-and-dirty-business.html

15. Miami-Dade County. *Trench Safety Act Compliance Statement, FM5238 Rev. (12-00).* Retrieved February 8, 2011 from http://facilities.dadeschools.net/form_pdfs/5238.pdf

Slide 8

16. New York City Department of Buildings. *Excavation and Trench Safety Guidelines* by Dan Eschenasy. www.NYC.gov/buildings. Retrieved February 8, 2011 from http://www.elcosh.org/en/document/161/d000168/trenching-is-a-dangerous-and-dirty-business.html

17. OSHA. *Working Safely in Trenches Safety Tips.* Retrieved February 8, 2011 from http://www.osha.gov/Publications/trench/trench_safety_tips_card.html

18. OSHA. *29CFR1926.650 subpart p. Excavations: scope, application and definitions.* Retrieved February 8, 2011 from http://www.osha.gov/pls/oshaweb/owadisp.show_document?p_id=10774&p_table=STANDARDS

19. OSHA. *29CFR1926.651 subpart p. Excavations: specific excavation requirements.* Retrieved February 8, 2011 from http://www.osha.gov/pls/oshaweb/owadisp.show_document?p_table=STANDARDS&p_id=10775

20. OSHA. *29CFR1926.652 subpart p. Excavations: requirements for protective systems.* Retrieved February 8, 2011 from http://www.osha.gov/pls/oshaweb/owadisp.show_document?p_table=STANDARDS&p_id=10776

21. OSHA. *OSHA Technical Manual SECTION V: CHAPTER 2 EXCAVATIONS: HAZARD RECOGNITON IN TRENCHING AND SHORING.* Retrieved February 8, 2011 from http://www.osha.gov/dts/osta/otm/otm_v/otm_v_2.html

22. OSHA. *OSHA's Construction e-tool.* Retrieved February 8, 2011 from http://www.osha.gov/SLTC/etools/construction/trenching/mainpage.html

The New York State Fatality Assessment and Control Evaluation (NY FACE) program is one of many workplace health and safety programs administered by the New York State Department of Health (NYSDOH). It is a research program designed to identify and study fatal occupational injuries. Under a cooperative agreement with the National Institute for Occupational Safety and Health (NIOSH), the NY FACE program collects information on occupational fatalities in New York State (excluding New York City) and targets specific types of fatalities for evaluation. NY FACE investigators evaluate information from multiple sources and summarize findings in narrative reports that include recommendations for preventing similar events in the future. These recommendations are distributed to employers, workers, and other organizations interested in promoting workplace safety. The NY FACE does not determine fault or legal liability associated with a fatal incident. Names of employers, victims and/or witnesses are not included in written investigative reports or other databases to protect the confidentiality of those who voluntarily participate in the program.

Additional information regarding the NY FACE program can be obtained from:
New York State Department of Health FACE Program
Bureau of Occupational Health
Flanigan Square, Room 230

1-518-402-7900
www.nyhealth.gov/nysdoh/face/face.htm

ACCIDENT REPORT

ACCIDENT SUMMARY No. 59

Accident Type:	Struck by Falling Wall	
Weather Conditions:	Clear/Wet Soil	
Type of Operation:	Trenching	
Size of Work Crew:	2	
Competent Safety Monitor on Site:	No	
Safety and Health Program in Effect:	Inadeqaute	
Was the Worksite Inspected Regularly:	No, short duration	
Training and Education Provided:	Some	
Employee Job Title:	Laborer	
Age & Sex:	27-Male	
Experience at this Type of Work:	1 Year	
Time on Project:	1 Day	

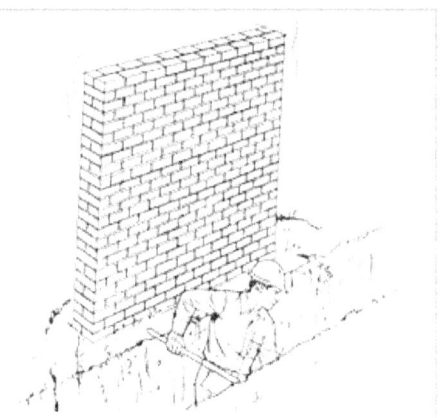

BRIEF DESCRIPTION OF ACCIDENT

An employee was in the process of locating an underground water line. A trench had been dug approximately 4 feet deep along side a brick wall 7 feet high and 5 feet long. The brick wall collapsed onto the victim who was standing in the trench. The injuries were fatal.

INSPECTION RESULTS

As a result of its investigation, OSHA issued citations for violation of the standard.

ACCIDENT PREVENTION RECOMMENDATIONS

The contractor should not permit employees to excavate below the level of the base of foundation footings when walls are unpinned [29 CFR 1926.651(i)(1)]

SOURCES OF HELP

- **OSHA 2202 Construction Industry Digest** ⁻ includes all OSHA construction standards and those general industry standards that apply to construction. Order No. 029-016-00151-4, ($2.25). Available from the Superintendent of Documents, Government Printing Office, Washington DC 20402-9325, phone (202) 512-1800. Make checks payable to Superintendent of Documents. For phone orders, Visa® or MasterCard®.
- **OSHA 2254 Training Requirements in OSHA Standards and Training Guidelines** ⁻ includes all OSHA construction standards and those general industry standards that apply to construction. Order No. 029-016-00160-3, ($6.00). Available from the Superintendent of Documents, Government Printing Office, Washington DC 20402-9325, phone (202) 512-1800. Make checks payable to Superintendent of Documents. For phone orders, Visa® or MasterCard®.
- **OSHA Safety and Health Guidelines for Construction** (Available from the National Information Service, 5285 Port Royal Road, Springfield, VA 22161; (703) 605-6000 or (800) 553-6847; Order No. PB-239-312/AS, $27). Guidelines to helpconstruction employers establish a training program in the safe use of equipment, tools, and machinery on the job.

- For information on OSHA-funded free consultation services call the nearest OSHA area office listed in telephone directories under U.S. Labor Department or under the state government section where states administer their own OSHA programs.
- Courses in construction safety are offered by the OSHA Training Institute, 1555 Times Drive, Des Plaines, IL 60018, 708/297-4810.
- OSHA Safety and Health Training Guidelines for Construction (Available from the National Technical Information Service, 5285 Port Royal Road, Springfield, VA 22161; 703/487-4650; Order No. PB-239-312/AS): guidelines to help construction employers establish a training program in the safe use of equipment, tools, and machinery on the Job.

NOTE: The case here described was selected as being representative of fatalities caused by improper work practices. No special emphasis or priority is implied nor is the case necessarily a recent occurrence. The legal aspects of the incident have been resolved, and the case is now closed.

Scaffolding Accidents

Design as a Risk Factor: Australian Study, 2000–2002

- Main finding: design contributes significantly to work-related serious injury

- 37% of workplace fatalities are due to design-related issues

- In another 14% of fatalities, design-related issues may have played a role

[Driscoll et al. 2008]

Photo courtesy of Thinkstock

Reinforced Concrete

NOTES

Several studies around the world have demonstrated that design can directly affect the safety of a construction site or process. The Australian government investigated the design-related root causes of their work-related fatalities. Seventy-seven (37%) of the 210 identified workplace fatalities definitely or probably had design-related issues involved. In another 29 fatalities (14%), the circumstances suggested that design issues were involved. The most common scenarios involved problems with rollover protective structures and/or associated seat belts; inadequate guarding; lack of residual current devices; inadequate fall protection; failed hydraulic lifting systems in vehicles and mobile equipment; and inadequate protection mechanisms on mobile plants and vehicles.

These fatal incidents might have been prevented if the hazards that caused them had been considered during the design phase.

SOURCES

Driscoll TR, Harrison JE, Bradley C, Newson RS [2008]. The role of design issues in work-related fatal injury in Australia. J Safety Res. *39*(2):209–14 [Epub 2008:Mar 13; PubMed index for MEDLINE: 18454972].

NIOSH Fatality Assessment and Control Evaluation (FACE) Program [1983]. Fatal incident summary report: scaffold collapse involving a painter. FACE 8306 [www.cdc.gov/niosh/face/In-house/full8306.html].

Photo courtesy of Thinkstock

Fatal Incident Summary Report: Scaffold Collapse Involving a Painter

INTRODUCTION

The National Institute for Occupational Safety and Health (NIOSH), Division of Safety Research (DSR), is currently conducting the Fatal Accident Circumstances and Epidemiology (FACE) Study. By scientifically collecting data from a sample of similar fatal accidents, this study will identify and rank factors which increase the risk of fatal injury for selected employees.

On May 25, 1983, a painter suffered fatal injuries when the suspended scaffolding from which he was working collapsed. The County Coroner requested NIOSH technical assistance to develop information on factors involved with the incident data.

CONTACTS/ACTIVITIES

After receiving notification, three Division of Safety Research personnel, a safety specialist, a safety engineer, and an epidemiologist, visited at the site to interview the employer and witnesses and to obtain comparison data from suitable co-workers. The research team, the police department, and the employer examined the impounded scaffold at an independent testing laboratory.

A debriefing session was held with the employer, other employees, and the contractor. During this introductory meeting, background information was obtained about the contractor and the employer, including an overview of their safety and health program. Interviews were conducted with witnesses and co-workers. Examining the scaffold assisted the researchers in developing hypotheses about the sequence of events leading to the incident.

SYNOPSIS OF EVENTS

The two workers had placed the scaffold supporting wire rope on the 7th floor permanently installed eye hooks. They then reeved the wire rope to the scaffold stirrups which are located at each end of the scaffold staging. After reeving was complete, the workers raised the scaffolding to the 7th floor windows. This action was accomplished by turning the drive motor directional switch to the "up" position and holding the motor switch in the "on" position.

The victim had to apply caulking around the windows. After caulking half way across the floor, he had to change positions, including independent life lines with a co-worker, who survived the incident. After caulking the remaining windows, the workers switched positions again in order to begin their descent.

The co-worker stated that he turned away from the victim and faced his stirrup in preparation of descent. As he did this, he felt some movement in the scaffold. He turned and looked at the victim, who motioned by hand signal to turn the directional switch to the "down" position. The co-worker signaled "okay" and turned to face his stirrup. As he was in the process of preparing his stirrup for downward movement plus getting his lanyard grab device ready to move down, he felt several sudden jerks and was suddenly dangling from his life line. After regaining his composure, the co-worker looked for the victim in the area of his life line. The co-worker then noticed the victim lying in the street across from the building.

GENERAL CONCLUSIONS AND RECOMMENDATIONS

There is some evidence which indicates the deceased was not familiar with the operation of this type of scaffold. For this type of scaffold, the operator must operate the drill and a brake lever at the same time with one hand, while releasing his lanyard on the safety line with the other hand.

Additionally, the victim's lanyard failed to prevent the fatal fall for one of two reasons. Either the lanyard was deteriorated to the extent that the impact load was in excess of the lanyard strength or the lanyard became entangled in the scaffold components.

It is suspected that the wire rope broke because the hoist's secondary safety mechanism did not function quickly enough. The wire rope broke at a level 20+ feet below where the scaffold was originally positioned. When the mechanism finally activated, the force of the falling scaffold caused the emergency braking cam to squeeze the rope to such an extent that it actually cut 5 of the 6 strands. The remaining strand was not of sufficient strength to hold the falling scaffold and it also broke.

It is recommended that workers who use scaffolds should be trained in the proper use, maintenance, and limitations of scaffolding, life lines and lanyards. Also management should be aware of their responsibilities when their workers are using scaffolds. Safety requirements for scaffolding are outlined in the OSHAct regulations 1910.28, 1910.29 and 1926.451.

Accidents Linked to Design

- 22% of 226 injuries that occurred from 2000 to 2002 in Oregon, Washington, and California were linked partly to design [Behm 2005]

- 42% of 224 fatalities in U.S. between 1990 and 2003 were linked to design [Behm 2005]

- In Europe, a 1991 study concluded that 60% of fatal accidents resulted in part from decisions made before site work began [European Foundation for the Improvement of Living and Working Conditions 1991]

- 63% of all fatalities and injuries could be attributed to design decisions or lack of planning [NOHSC 2001]

NOTES

Research conducted in the United States, Europe, and other regions has shown that design does affect the inherent risk in constructing a facility. Research linked design to 22% of injuries that occurred in western states and 42% of fatalities across the country. European researchers found that nearly two-thirds of fatalities and injuries were linked to design. Facility designers are encouraged to consult with occupational safety and health professionals early in the design process to identify and design out hazards and to reduce risk of injury, illness, and death.

SOURCES

Behm M [2005]. Linking construction fatalities to the design for construction safety concept. Safety Sci 43:589–611.

NOHSC [2001]. CHAIR safety in design tool. New South Wales, Australia: National Occupational Health & Safety Commission.

European Foundation for the Improvement of Living and Working Conditions [1991]. From drawing board to building site (EF/88/17/FR). Dublin: European Foundation for the Improvement of Living and Working Conditions.

Falls

 **Falls**

- Number one cause of construction fatalities
  - in 2010, 35% of 751 deaths
    www.bls.gov/news.release/cfoi.t02.htm

- Common situations include making connections, walking on beams or near openings such as floors or windows

- Fall protection is required at height of 6 feet above a surface [29 CFR 1926.760].

- Common causes: slippery surfaces, unexpected vibrations, misalignment, and unexpected loads

Reinforced Concrete

NOTES

Falls are the number one cause of deaths in the construction industry. In 2004, 445 (36%) of 1,234 deaths were due to falls [BLS 2006]. By contrast, of 751 deaths in the construction sector in 2010, 35% were attributed to falls [BLS 2011a]. The decline in number of fatalities in the construction sector in 2010, compared to 2004, was attributed more to the economic downturn than to any other factor, however [BLS 2011b].

Falls from any height can be fatal. In construction, workers are often high off the ground. For structural reasons, the taller cross-sections of W shapes are usually chosen for beams. The flanges on W shapes may be less than six inches wide. Workers walk on beams, sometimes without fall protection. Fall protection is highly recommended and often required in most scenarios involving heights. OSHA requires fall protection at a height of 15 feet above a surface during steel erection. For other construction phases, it is 6 feet [29 CFR 1926.760].

SOURCES

BLS [2011a]. Census of Fatal Occupational Injuries. Washington, DC: U.S. Department of Labor, Bureau of Labor Statistics [www.bls.gov/news.release/cfoi.t02.htm].

BLS [2011b]. Injuries, Illnesses, and Fatalities (IIF). Washington, DC: U.S. Department of Labor, Bureau of Labor Statistics [www.bls.gov/iif/home.htm].

BLS [2006]. Injuries, illnesses, and fatalities in construction, 2004. By Meyer SW, Pegula SM. Washington, DC: U.S. Department of Labor, Bureau of Labor Statistics, Office of Safety, Health, and Working Conditions [www.bls.gov/opub/cwc/sh20060519ar01p1.htm].

OSHA [2001]. Standard number 1926.760: fall protection. Washington, DC: U.S. Department of Labor, Occupational Safety and Health Administration.

 **Death from Injury**

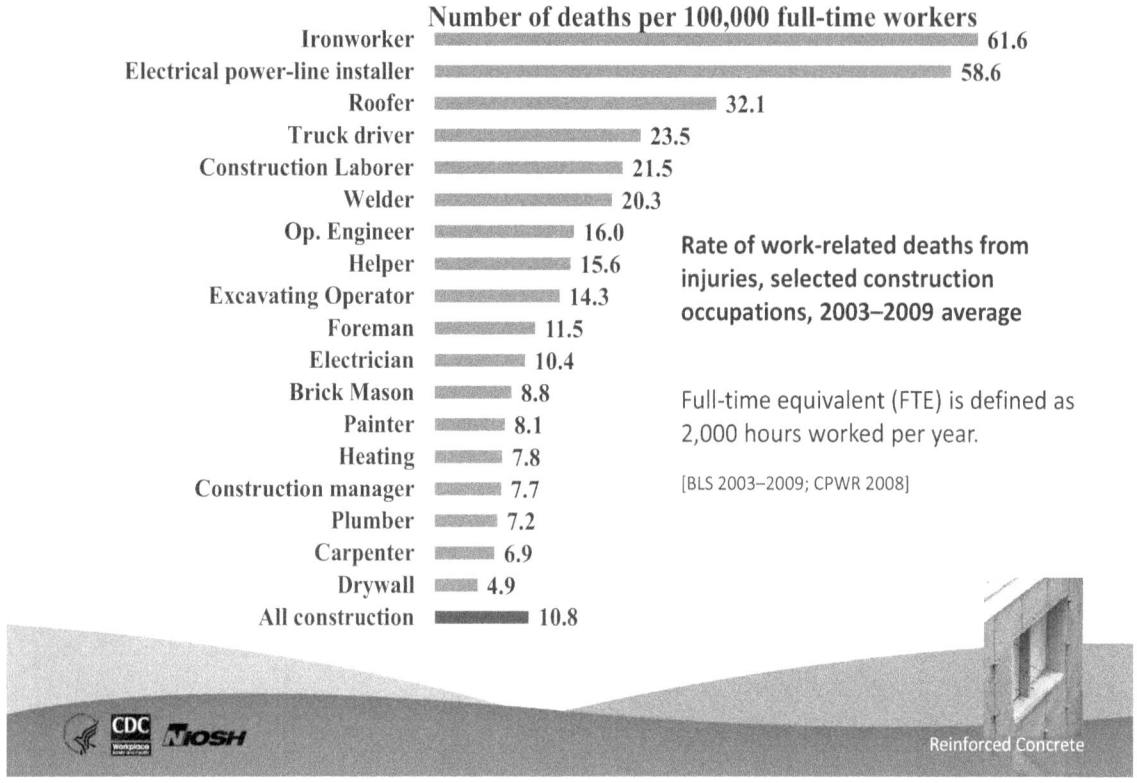

Number of deaths per 100,000 full-time workers

Occupation	Rate
Ironworker	61.6
Electrical power-line installer	58.6
Roofer	32.1
Truck driver	23.5
Construction Laborer	21.5
Welder	20.3
Op. Engineer	16.0
Helper	15.6
Excavating Operator	14.3
Foreman	11.5
Electrician	10.4
Brick Mason	8.8
Painter	8.1
Heating	7.8
Construction manager	7.7
Plumber	7.2
Carpenter	6.9
Drywall	4.9
All construction	10.8

Rate of work-related deaths from injuries, selected construction occupations, 2003–2009 average

Full-time equivalent (FTE) is defined as 2,000 hours worked per year.

[BLS 2003–2009; CPWR 2008]

Reinforced Concrete

NOTES

The Center for Construction Research and Training compiles a "Construction Chart Book" using Bureau of Labor Statistics data [CPWR 2008]. It includes two illuminating charts useful for considering safety issues. This chart is compiled from 2003–2009 data on workplace fatalities. Ironworkers experience the highest work-related death rate, with 61.6 fatalities per 100,000 FTE.

SOURCES

BLS [2003–2009]. Census of Fatal Occupational Injuries. Washington, DC: U.S. Department of Labor, Bureau of Labor Statistics [www.bls.gov/iif/oshcfoi1.htm].

CPWR [2008]. The construction chart book. 4th ed. Silver Spring, MD: Center for Construction Research and Training.

Fatality Assessment and Control Evaluation

NIOSH FACE Program www.cdc.gov/niosh/face

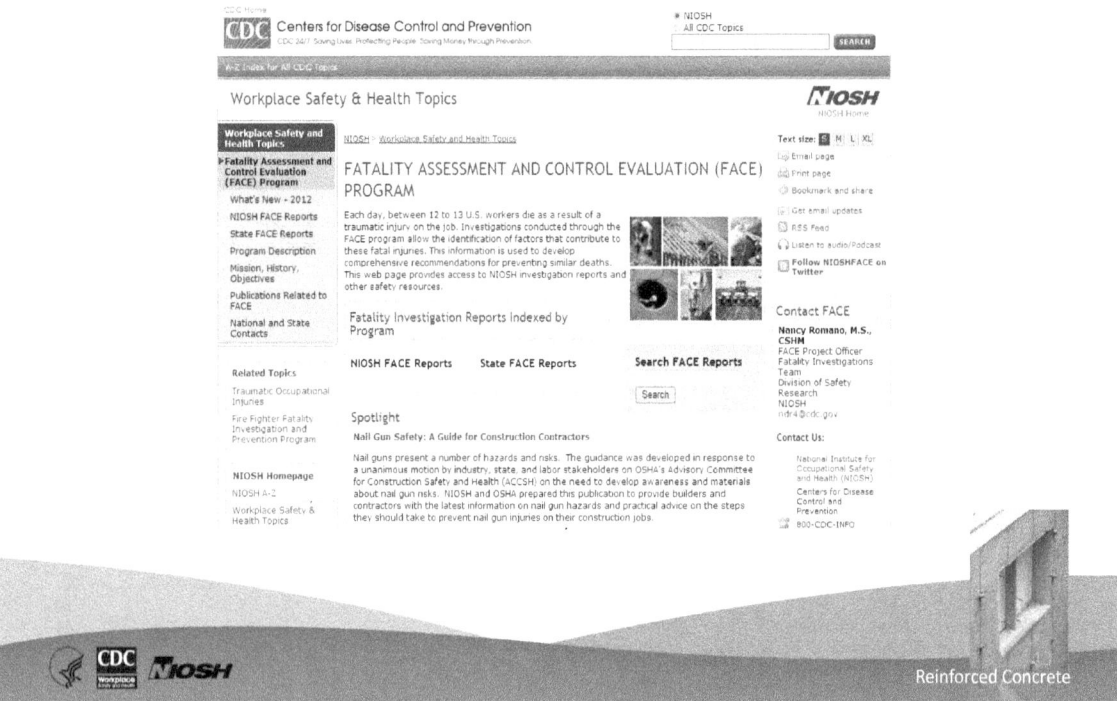

NOTES

The NIOSH Fatality Assessment and Control Evaluation Program examines worker fatalities by type of injury. By studying these reports, an enterprising designer can identify recurrent problems to "design out."

SOURCE

NIOSH Fatality Assessment and Control Evaluation Program [www.cdc.gov/niosh/face/]

 **What is Prevention through Design?**

Eliminating or reducing work-related hazards and illnesses and minimizing risks associated with

- Construction

- Manufacturing

- Maintenance

- Use, reuse, and disposal of facilities, materials, and equipment

NOTES

PtD is a risk management technique that is being applied successfully in many industries, including manufacturing, healthcare, telecommunications, and construction. PtD is the optimal method of preventing occupational illnesses, injuries, and fatalities by designing out the hazards and risks. This approach involves the design of tools, equipment, systems, work processes, and facilities in order to reduce, or eliminate, hazards associated with work. The concept is simply that the safety and health of workers throughout the life cycle are considered while the product and/or process is being designed. The life cycle starts with concept development, and includes design, construction or manufacturing, operations, maintenance, and eventual disposal of whatever is being designed, which could be a facility, a material, or a piece of equipment.

PtD processes have been required in other countries for several years now, but in the United States, PtD is being adopted on a voluntary basis. The National Institute for Occupational Safety and Health (NIOSH) is spearheading a national initiative in PtD and partnering with many professional organizations to apply the concept to their industry and professions. The Occupational Safety and Health Administration (OSHA) is very interested in PtD but is not currently considering making it mandatory.

PtD design professionals (that is, architects and/or engineers) working with the project owner (that is, the client) make deliberate design decisions that eliminate or reduce the risk of injuries or illness throughout the life of a project, beginning at the earliest stages of a project's life cycle. PtD is thus the deliberate consideration of construction and maintenance worker safety and health in the design phase of a construction project. PtD processes in construction have been required in the United Kingdom for over a decade and are being implemented in other countries such as Australia and Singapore.

PtD applies to the design of a facility, that is, to the aspects of the completed building that make a project inherently safer. PtD does not focus on how to make different methods of construction safer. For example, it does not focus on how to use fall protection systems, but it does include consideration of design decisions that influence how often fall protection will be needed. Similarly, PtD does not address how to erect safe scaffolding, but it does relate to design decisions that influence the location and type of scaffolding needed to accomplish the work. PtD concepts may also be used to design temporary structures. Some design decisions improve workplace safety. For example, when the height of parapet walls is designed to be 42", the parapet acts as a guardrail and enhances safety. When designed into the permanent structure of the building and sequenced early in construction, the parapet at this height acts to enhance safety during initial construction activities and during subsequent maintenance and construction activities, such as roof repair. In the United States, the employer is solely responsible for site safety.

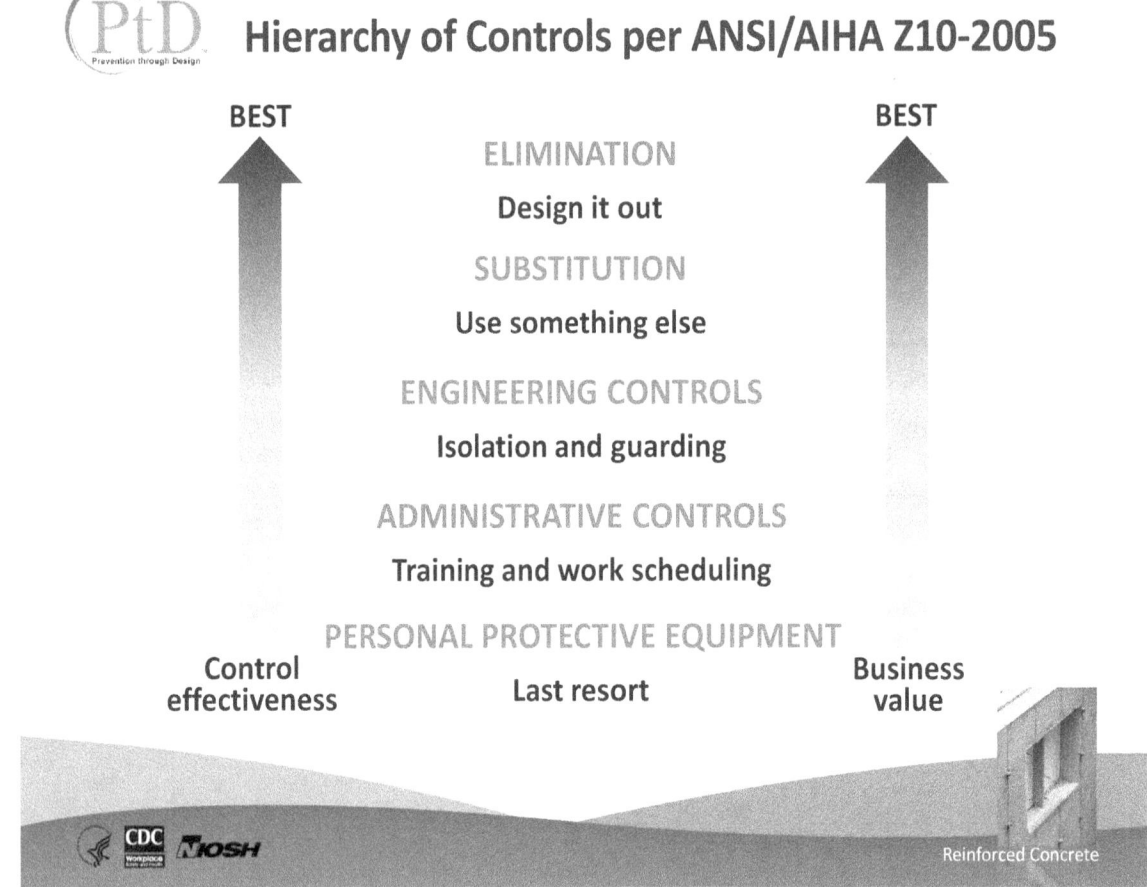

Hierarchy of Controls per ANSI/AIHA Z10-2005

BEST   BEST

ELIMINATION

Design it out

SUBSTITUTION

Use something else

ENGINEERING CONTROLS

Isolation and guarding

ADMINISTRATIVE CONTROLS

Training and work scheduling

PERSONAL PROTECTIVE EQUIPMENT

Control effectiveness   **Last resort**   **Business value**

Reinforced Concrete

NOTES

This slide shows the well-accepted Hierarchy of Controls. PtD anticipates and removes potential hazardous elements at the design phase of a project through elimination or substitution. Residual risks may be minimized through the use of engineering and administrative controls.

The top of the hierarchy is better in terms of improved occupational safety and health (OSH) and cost savings. Below is a description of the different levels, from most to least effective.

Elimination: "Design out" hazards and hazardous exposures.

Substitution: Substitute less-hazardous materials, processes, operations, or equipment. A larger crane may be specified when the load or the reach approaches the crane design limit. Nontoxic chemicals are preferred. The Green Chemistry movement replaces toxic compounds with less hazardous chemicals.

Engineering controls: Isolate process or equipment or contain the hazard. Remove hazard from work zone, e.g., with exhaust ventilation. Require two hands to operate machinery. Use warning devices to warn worker about entry into hazard zone. Signs, labels, alarms, and flashing lights give warnings. Safety switches, hand guards, and other engineering controls prevent certain kinds of injuries.

Administrative controls: Job rotation, work scheduling, training, well-designed work methods, and organization are examples. Administrative controls include training modules and company procedures. A well-organized worksite is safer than a messy one. Reducing the clutter on a construction site improves worker safety by reducing the exposure to hazards. The foreman controls site layout and housekeeping policies.

Personal Protective Equipment (PPE): Includes but is not limited to safety glasses for eye protection; ear plugs for hearing protection; clothing such as safety shoes, gloves, and overalls; face shields for welders; fall harnesses; and respirators to prevent inhalation of hazardous substances.

SOURCE

ANSI/AIHA [2005]. American national standard for occupational health and safety management systems. New York: American National Standards Institute, Inc. ANSI/AIHA Z10-2005.

Personal Protective Equipment (PPE)

- Last line of defense against injury
- Examples:
  - Hard hats
  - Steel-toed boots
  - Safety glasses
  - Gloves
  - Harnesses

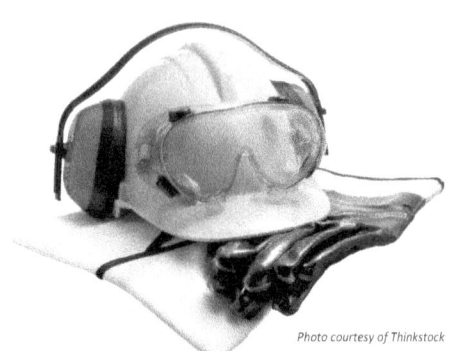

Photo courtesy of Thinkstock

OSHA www.osha.gov/Publications/osha3151.html

Reinforced Concrete

NOTES

Personal Protective Equipment, or PPE, includes items worn as a last line of defense against injury. OSHA-required PPE can include hardhats, steel-toed boots, safety glasses or safety goggles, gloves, earmuffs, full body suits, respiratory aids, face shields, and fall harnesses.

SOURCES

NOHSC [2001]. CHAIR safety in design tool. New South Wales, Australia: National Occupational Health & Safety Commission.

OSHA PPE publications

www.osha.gov/Publications/osha3151.html

www.osha.gov/OshDoc/data_General_Facts/ppe-factsheet.pdf

www.osha.gov/OshDoc/data_Hurricane_Facts/construction_ppe.pdf

Photo courtesy of Thinkstock

51

PtD Process

[Hecker et al. 2005]

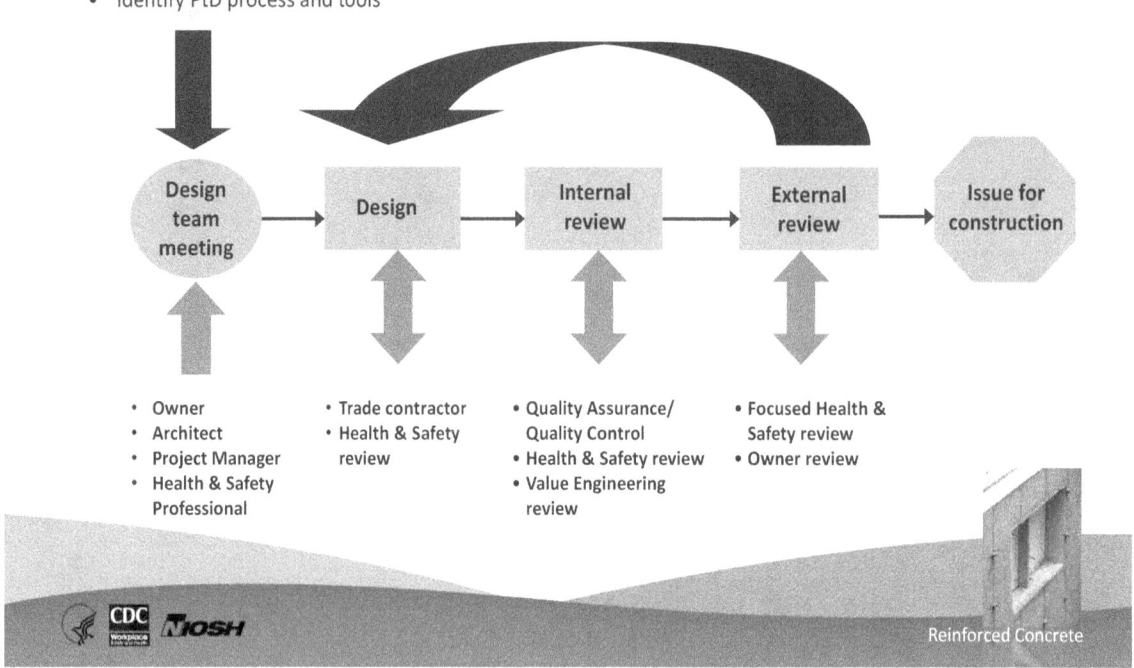

- Establish PtD expectations
- Include construction and operation perspective
- Identify PtD process and tools

Design team meeting → Design → Internal review → External review → Issue for construction

- Owner
- Architect
- Project Manager
- Health & Safety Professional

- Trade contractor
- Health & Safety review

- Quality Assurance/ Quality Control
- Health & Safety review
- Value Engineering review

- Focused Health & Safety review
- Owner review

Reinforced Concrete

NOTES

This graphic depicts the typical PtD process. The key component of this process is the incorporation of safety knowledge into design decisions. For example, site safety should be considered throughout the design process. A progress review specifically focused on site safety may be effective. Site safety knowledge can be provided by trade contractors, an on-site employee, or a hired consultant. The graphic emphasizes the importance of communication between designers and constructors. Such communication during design may reveal steps to reduce construction duration.

Many Project Managers schedule a Value Engineering review prior to issuing drawings for bid. The purpose is to reduce overall project costs. Unfortunately, during the review, redundant systems that are necessary to protect worker health may be eliminated. It is therefore considered a best practice to conduct a focused Health & Safety (H&S) review before drawings are issued.

SOURCE

Hecker S, Gambatese J, Weinstein M [2005]. Designing for worker safety: moving the construction safety process upstream. Prof Saf 50(9):32–44.

Integrating Occupational Safety and Health with the Design Process

Stage	Activities
Conceptual design	Establish occupational safety and health goals, identify occupational hazards
Preliminary design	Eliminate hazards, if possible; substitute less hazardous agents/processes; establish risk minimization targets for remaining hazards; assess risk; and develop risk control alternatives. Write contract specifications.
Detailed design	Select controls; conduct process hazard reviews
Procurement	Develop equipment specifications and include in procurements; develop "checks and tests" for factory acceptance testing and commissioning
Construction	Ensure construction site safety and contractor safety
Commissioning	Conduct "checks and tests," including factory acceptance; pre–start up safety reviews; development of standard operating procedures (SOPs); risk/exposure assessment; and management of residual risks
Start up and occupancy	Educate; manage changes; modify SOPs

NOTES

The integration of OSH goals within the design processes is an essential concept because it elevates the importance of safety and health as a value proposition in the overall design, construction, and operation of projects.

Identify hazards during conceptual design. Follow the Hierarchy of Controls to eliminate or reduce risks.

For example, how much space is needed to access, maintain, and replace HVAC units?

Use project specifications to require the inclusion of fall protection systems such as permanent anchor points for lifelines. Reduce fall hazards by specifying a ladder-free construction site.

Obtain a site plan that shows the location of existing underground and overhead utilities and develop traffic control plans to avoid those hazards.

Compare the list of desirable safety features against the detailed design.

Obtain feedback from H&S professionals, contractors, and trade representatives. Modify the design to improve safety.

Call out required hazard controls on the drawing and in the contract specifications when possible. During procurement, compare materials and equipment received against the contract specifications. Develop a checklist for commissioning.

During construction, how do contractors communicate with the project manager and each other? Who has the authority to correct a hazardous condition on the worksite?

What procedures are followed before and after permanent equipment reaches the site? Follow the commissioning checklist!

- Does the building have unusual features? Educate the owners and tenants.
- Are special operating procedures required?
- At each stage of the design process, think of ways to reduce the workplace risks.

Safety Payoff During Design

[Adapted from Szymberski 1997]

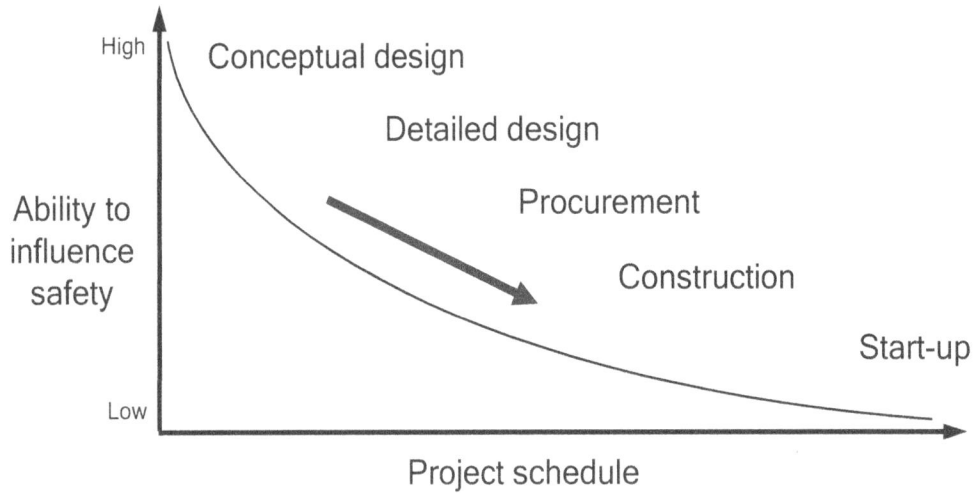

NOTES

Most owners and design professionals know intuitively that the earlier in the design process that cost is considered, the easier it is to achieve cost-effective goals. The same is true for construction duration and quality. A worker's ability to influence project criteria decreases as the design and construction progress. The same principle is true for construction safety. The earlier in the project life cycle that safety is considered, the easier it is to reduce hazards. This concept is in contrast to the prevailing methods of planning for construction site safety, which do not begin until a short time before the construction phase, when the ability to influence safety is limited.

SOURCE

Szymberski R [1997]. Construction project planning. TAPPI J *80*(11):69–74.

PtD Process Tasks

[Adapted from Toole 2005; Hinze and Wiegand 1992]

- Perform a hazard analysis

- Incorporate safety into the design documents

- Make a CAD model for member labeling and erection sequencing

Photo courtesy of Thinkstock

Reinforced Concrete

NOTES

This slide provides more details about the PtD process. Before, during, or after the conceptual design of a building, a hazard analysis can be performed. The designer meets with field professionals to review constructability, looking through the entire design for any hazards and addressing those hazards. The field professional can teach an inexperienced designer how to minimize risks in the field.

The safety input received during conceptual design can be reflected in detailed design drawings and specifications. Another constructability review should occur as the detailed design nears completion.

Sometimes the drawings that result from a PtD process look the same as typical construction drawings, but they are inherently safer for construction. Other times, drawings include special details and labels to make it easier for workers to erect the design safely.

Construction documents can be supplemented with graphic models and tables that contribute to safe erection. For example, a CAD file can be used to label steel members for safe erection sequencing. New software such as building information modeling (BIM) is able to show the final layouts of buildings and can detect any spatial problems before construction starts. Clearly labeled shop drawings eliminate confusion during installation. The BIM program can recommend efficient, safer erection sequencing.

SOURCES

Hinze J, Wiegand F [1992]. Role of designers in construction worker safety. Journal of Construction Engineering and Management *118*(4):677–684.

Toole TM [2005]. Increasing engineers' role in construction safety: opportunities and barriers. Journal of Professional Issues in Engineering Education and Practice *131*(3):199–207.

Photo courtesy of Thinkstock

 **Designer Tools**

- Checklists for construction safety [Main and Ward 1992]
- Design for construction safety toolbox [Gambatese et al. 1997]
- Construction safety tools from the UK or Australia
  - Construction Hazard Assessment Implication Review (CHAIR) [NOHSC 2001]

NOTES

Most designers are not trained in PtD or construction site safety. It is therefore critical that they be given tools to facilitate the process. A PtD checklist alerts designers to common design elements that can lead to unnecessary hazards and identifies design options that are inherently safer. An example checklist is provided on the next slide.

The Design for Construction Safety Toolbox was developed by a Construction Industry Institute–sponsored research team that included leading PtD academics. This Toolbox was recently updated by Professor Jimmie Hinze at the University of Florida. The United Kingdom and Australia make available on the Web valuable PtD tools that reflect their experiences with PtD legislation and voluntary initiatives. For example, CHAIR (Construction Hazard Assessment Implication Review) is an Australian tool and methodology that systematically combines brainstorming and decisions to gradually rid the design of unnecessary hazards.

SOURCES

NOHSC [2001]. CHAIR safety in design tool. New South Wales, Australia: National Occupational Health & Safety Commission.

Gambatese JA, Hinze J, Haas CT [1997]. Tool to design for construction worker safety. J Arch Eng 3(1):2–41.

Main BW, Ward AC [1992]. What do engineers really know and do about safety? Implications for education, training, and practice. Mechanical Engineering 114(8):44–51.

Example Checklist

Item	Description
1.0	**Structural Framing**
1.1	Space slab and mat foundation top reinforcing steel at no more than 6 inches on center each way to provide a safe walking surface.
1.2	Design floor perimeter beams and beams above floor openings to support lanyards.
1.3	Design steel columns with holes at 21 and 42 inches above the floor level to support guardrail cables.
2.0	**Accessibility**
2.1	Provide adequate access to all valves and controls.
2.2	Orient equipment and controls so that they do not obstruct walkways and work areas.
2.3	Locate shutoff valves and switches in sight of the equipment which they control.
2.4	Provide adequate head room for access to equipment, electrical panels, and storage areas.
2.5	Design welded connections such that the weld locations can be safely accessed.

[Checklist courtesy of John Gambatese]

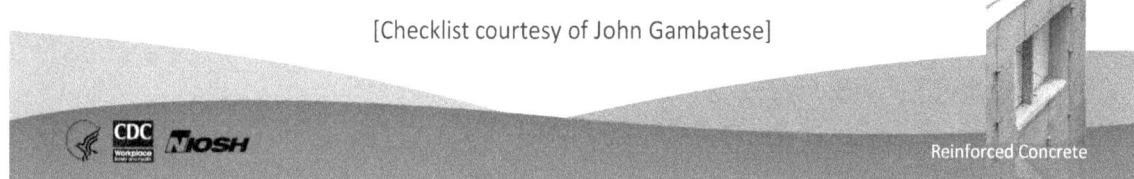

Reinforced Concrete

NOTES

Like many PtD checklists, this example includes hazards associated with both construction and maintenance.

SOURCE

Checklist courtesy of John Gambatese

 **OSHA silica eTool**

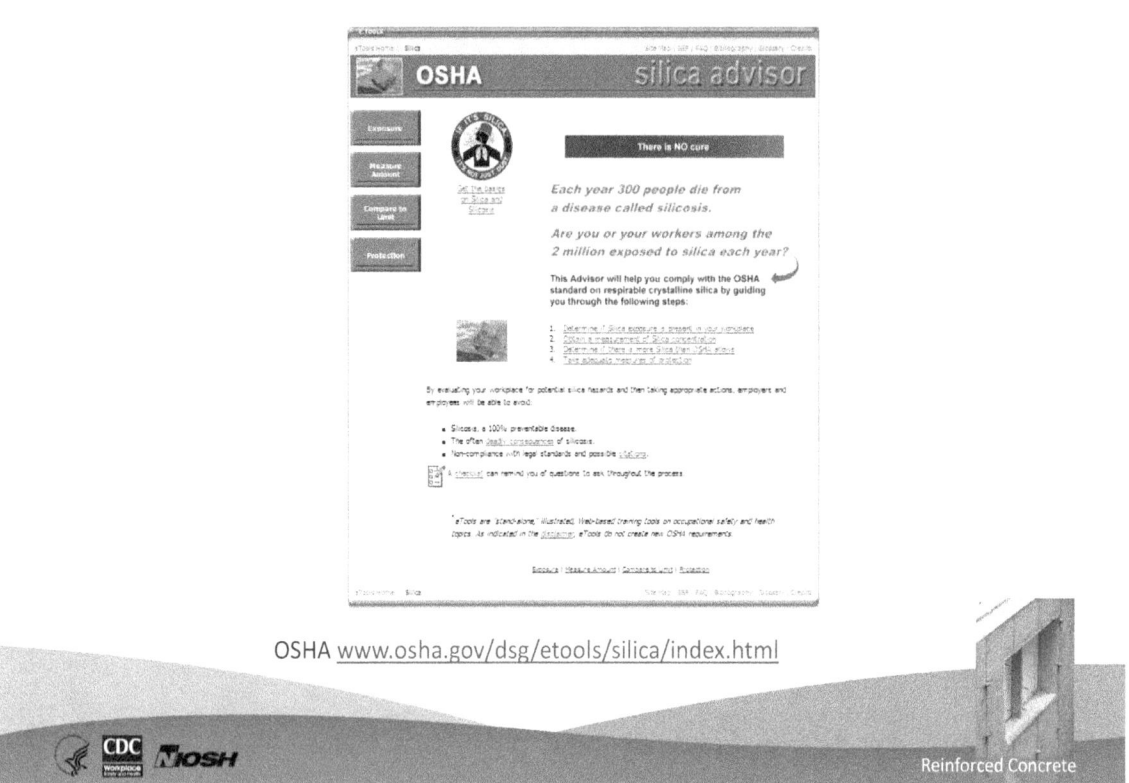

OSHA www.osha.gov/dsg/etools/silica/index.html

Reinforced Concrete

NOTES

To help concrete construction employers determine what their employee safety responsibilities are, OSHA maintains Web pages that contain interpretations and clarifications of the federal standards and provide access to eTools that may be adopted on a voluntary basis. This slide shows the silica eTool, which helps communicate the requirements for preventing silicosis, a common deadly disease caused by breathing dust containing silica particles. Silica dust is often generated, sometimes at quite high concentrations, during concrete construction, maintenance, and demolition operations.

SOURCE

OSHA [www.osha.gov/dsg/etools/silica/index.html]

Why Prevention through Design?

- Ethical reasons

- Construction dangers

- Design-related safety issues

- Financial and non-financial benefits

- Practical benefits

Photo courtesy of Thinkstock

Reinforced Concrete

NOTES

Engineers have strong ethical reasons to apply the PtD concept to their designs. There are practical benefits, too. Lost-time accidents delay the job, destroy crew morale, and cost money. The next few slides will show there are many reasons why owners and design professionals should be motivated to incorporate PtD in a project.

SOURCE

Photo courtesy of Thinkstock

Ethical Reasons for PtD

- National Society of Professional Engineers' Code of Ethics:

   "Engineers shall hold paramount the safety, health, and welfare of the public..."

- American Society of Civil Engineers' Code of Ethics:

   "Engineers shall recognize that the lives, safety, health and welfare of the general public are dependent upon engineering decisions..."

NSPE www.nspe.org/ethics

ASCE www.asce.org/content.aspx?id=7231

NOTES

Many safety professionals and design professionals believe that PtD is clearly an ethical duty. Nearly all national engineering societies include in their code of ethics a statement similar to the one shown here for the National Society of Professional Engineers: "Engineers shall hold paramount the safety, health, and welfare of the public."

The American Society of Civil Engineers goes one step further and explicitly states that engineering decisions directly affect safety. These organizations pledge to protect the public. Why? The public lacks the knowledge of forces, stresses, and other risk-related issues that contribute to hazardous work-related conditions. Many construction and maintenance workers, especially apprentices, fail to perceive an unsafe condition. Even if construction workers recognize a hazard that could have been eliminated or reduced through an alternative design, there are significant barriers to redesign after construction is under way. Their safety and health deserve consideration.

SOURCES

American Society of Civil Engineers [ASCE] [www.asce.org/Content.aspx?id=7231]

National Society of Professional Engineers [NSPE][www.nspe.org/ethics]

PtD Applies to Constructability

- How reasonable is the design?
  - Cost
  - Duration
  - Quality
  - Safety

Photo courtesy of the Cincinnati Museum Center www.cincymuseum.org

Reinforced Concrete

NOTES

Most designers know that what may look great on paper might not be constructible. An important part of the design process is to evaluate the design's constructability, that is, to what extent the design can be constructed at a reasonable price, quickly, and with high quality. Safety is an important part of constructability. Accidents cost money, delay construction, and may result in bad publicity rather than acclaim for the owner.

Exciting buildings designed by creative architects require strong consideration of worker safety and health early in the design process. Owners realize these one-of-a-kind structures cost more to build and generally present unique challenges for the construction crew. Fewer construction firms have the expertise needed to build the structure, so fewer firms submit a bid, which reduces competition and therefore drives up price, resulting in higher bond and insurance costs. The timeline for procurement and construction is harder to estimate. The uniqueness of the design creates construction and maintenance challenges. Unusual materials, custom fabrications, non-standard specifications, and striking aesthetic features inherent in these designs require greater collaboration. The PtD process shown on the next slide helps the design team identify potential hazards in time to devise appropriate prevention strategies for construction crews and future maintenance workers. The project manager should include occupational safety and health professionals throughout the design process to design-in protections for workers.

SOURCE
Photo courtesy of the Cincinnati Museum Center

 **Business Value of PtD**

- Anticipate worker exposures—be proactive
- Align health and safety goals with business goals
- Modify designs to reduce/eliminate workplace hazards in

Facilities	Equipment
Tools	Processes
Products	Work flows

 Improve business profitability!

AIHA www.ihvalue.org

Reinforced Concrete

NOTES

Companies that have implemented PtD programs experience lower than average injury and illness rates and lower workers' compensation expenses. However, the business value of PtD does not end there. In a study entitled Demonstrating the Business Value of Industrial Hygiene (known as The Value Study), findings showed that significant business cost savings accrue when hazards are eliminated or reduced.

SOURCE

American Institute of Industrial Hygienists [AIHA] [2008]. Strategy to demonstrate the value of industrial hygiene [www.aiha.org/votp_NEW/pdf/votp_exec_summary.pdf].

Benefits of PtD

- Reduced site hazards and thus fewer injuries
- Reduced workers' compensation insurance costs
- Increased productivity
- Fewer delays due to accidents
- Increased designer-constructor collaboration
- Reduced absenteeism
- Improved morale
- Reduced employee turnover

NOTES

PtD yields better value for owners and better health for the workers. When a project is designed with construction worker safety in mind, there are fewer hazards on site, with fewer injuries and fatalities. A reduction in injuries results in reduced workers' compensation insurance and less down-time, a direct savings for the employer. Experience shows PtD increases productivity and reduces labor costs. Safer designs lead to fewer project delays.

Industries Use PtD Successfully

- Construction companies
- Computer and communications corporations
- Design-build contractors
- Electrical power providers
- Engineering consulting firms
- Oil and gas industries
- Water utilities

  And many others

NOTES

Major corporations in diverse industries and public utilities in several states have applied PtD through initiatives or established programs. At these companies, worker safety and health are an integral part of the corporate culture. International construction firms first encountered PtD on their European projects. They brought the concepts and related cost savings home to their American operations. Many firms provide PtD training for their design engineers in the areas of construction site safety, PtD checklists, and safety constructability reviews. These firms want to hire engineers who have a basic understanding of PtD.

Elements, Activities, and Hazards

REINFORCED CONCRETE DESIGN

Elements, Activities, and Hazards

NOTES

Let's continue with a look at reinforced concrete design.

 **Introduction to Reinforced Concrete**

Topic	Slides
Elements	32–40
Design Process	41
Construction Activities	42
Construction Hazards	43–44

Structural Collapses During Construction

Adobe Acrobat Document

Structure Magazine www.structuremag.org/article.aspx?articleID=1177

Reinforced Concrete

NOTES

In this next section of slides, we will provide basic information about reinforced concrete design and construction, along with the hazards associated with reinforced concrete construction activities. One hazard shared by all reinforced concrete elements is the reliance on forms to support the structure until the concrete attains sufficient strength. Structural collapse occurs when construction loads exceed the design loads. Furthermore, buildings may be built to codes for final use that involve design loads which are much lower than the loads that can be created during the construction process from the presence of heavy construction equipment, materials, workers, etc. Shoring is needed to support poured concrete until it is hardened. In the article from STRUCTURE magazine, 20% of structural collapses studied were attributed to design errors (relating to permanent and temporary structures) and 80% were due to construction errors.

SOURCE

Ayub M [2010]. Structural collapses during construction: lessons learned, 1990–2008. Structure (December) [www.structuremag.org/article.aspx?articleID=1177].

Structural Collapses During Construction

Lessons Learned, 1990-2008

By Mohammad Ayub, P.E., S.E.

The Occupational Safety and Health Administration (OSHA) investigated 96 structural collapses during construction involving fatalities and injuries from 1990 to 2008. The most probable causes of these incidents are summarized in *Table 1* (available in the online version of this article; visit **www.STRUCTUREmag.org**). These incidents took the lives of 117 construction employees and caused injuries to another 235. The incidents occurred in a wide range of structures – steel, concrete and timber; high-rises and low-rises. The aggregate number of construction deaths due to all causes is staggering – approximately 1,000 in 2008 alone. *Figures 1* and *2* show the number of deaths and the rate of fatalities in the construction industry. As can be seen, the highest rate of deaths and injuries occurs in construction activities.

Construction errors contributed to 80% of the structural collapses investigated by OSHA. The remaining 20% of the incidents are attributed to structural design flaws on the part of either the structural engineer of record (SER) or a structural engineer retained by a contractor to design specific members. Steel structures, including scaffolds and platforms, were involved in 62% of these incidents.

Discussion

The largest group of structural collapses involved 60 steel structures:
- 14 structural steel frames
- 14 scaffolds
- 18 special steel structures and cranes
- 5 television antenna towers
- 3 cofferdams
- 6 steel roof trusses and joists

The second largest group involved 29 concrete and masonry structures:
- 3 concrete frames
- 12 shorings supporting freshly placed concrete
- 4 demolitions involving concrete structures
- 5 precast concrete structures
- 5 masonry walls

The third group consisted of wood structures:
- 7 wood frames and roof trusses

Construction Errors

1) In 47 cases, contractors did not generally follow the installation procedures prescribed and recommended by the manufacturers and designers, such as providing temporary bracing, lateral bracing, diagonal bracing, bridging and anchoring, guy cables, lateral supports, and proper welded connections.

2) In 15 cases, contractors overloaded certain structural members beyond their ultimate capacities.

3) In 9 cases, contractors did not provide temporary bracing during construction of steel frames, and concrete or masonry walls. As a result, wind pressures caused their collapse.

4) In 7 cases, contractors began to demolish existing structures without regard to structural stability and capacity of existing structural members.

Structural Design Errors

Out of 96 incidents, 19 construction incidents were related to structural design errors. These occurred in 13 steel structures, five concrete structures, and one masonry structure. 17 of these incidents are briefly described below.

Incident No. 4
Precast Concrete Beams

Two critical bottom reinforcing bars of a precast beam were not provided with the required development lengths. This resulted in a significantly reduced flexural strength of the beam, and hence the failure.

Lessons learned: The precast beam designer and detailer must indicate in their detail drawings the required rebar development lengths, including the rebar splice lengths per the ACI code.

Incident No. 6
Soldier Beam and Lagging Cofferdam

The installed outlookers between the soldier beams and walers did not have sufficient strength to resist the unbalanced lateral earth pressure.

Lessons learned: Outlookers that transfer forces from the waler to the soldier beam could be subjected to flexural stresses due to unbalanced earth pressure. Such outlookers must be designed to resist all anticipated forces.

Incident No. 8
Steel Stack

During the design of the steel stack, the SER did not consider vortex shedding under sustained wind speed.

Lessons learned: Uniform winds with little turbulence are known to create vortex shedding, which causes large vibrations in the across-wind direction in tall stacks of circular cross-sections under the condition of resonance. Winds that are not in a steady state do not create vortex shedding. The transverse resonance occurs when the shedding frequency becomes close to the natural frequency of the steel stack. The SER must consider a suitable abatement method.

Incident No. 15
Steel Sheeting Cofferdam

The engineer under-proportioned the depth of embedment of the cofferdam steel sheet piles, resulting in a "quick condition" and subsequent soil failure.

Lessons learned: Appropriate seepage forces at the bottom of the excavation must be considered in determining the depth of sheet piles.

Incident No. 16
Concrete Building

The formwork design engineer under-proportioned the support system for all reasonably anticipated vertical and lateral loads imposed on the formwork.

Lessons learned: The formwork design engineer must consider all anticipated vertical and lateral loads to be imposed on the formwork and proper load transfer to the base, and must provide detail drawings.

Incident No. 22
Steel Canopy Structure

The structural engineer did not properly design the canopy structure for the loads that were placed on it.

LESSONS LEARNED
problems and solutions encountered by practicing structural engineers

Work Related Fatalities

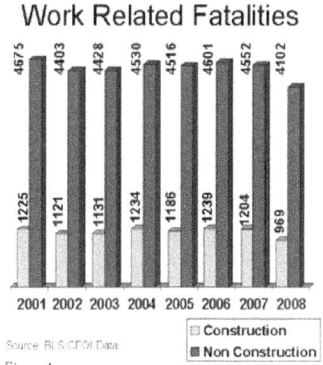

2001 2002 2003 2004 2005 2006 2007 2008

Source: BLS CFOI Data

□ Construction
■ Non Construction

Figure 1.

Fatality Rates

(Fatalities per 100,000 workers)

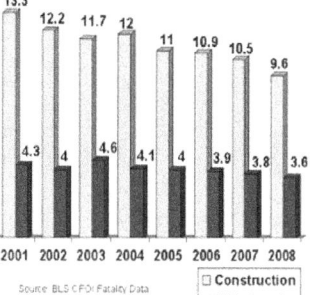

2001 2002 2003 2004 2005 2006 2007 2008

Source: BLS CFOI Fatality Data

□ Construction
■ All Industries

Figure 2.

Lessons learned: Final design should not be based on preliminary loads assumed during the initial design. Final design should be based on the anticipated final loads.

Incident No. 27
Masonry Foundation Wall

The structural engineer did not design a braced masonry foundation wall, and did not provide vertical reinforcement to resist the lateral loads.

Lessons learned: Unreinforced masonry walls are highly susceptible to overturning, especially during construction, unless they are braced. Bracing must stay in place until floor or roof framing installation is completed.

Incident No. 33
Bridge

The structural engineer under-proportioned the steel bracket supporting the steel beams of the catching platform under the center span of the bridge where the deck was being demolished. Also, the epoxy anchors used to support the brackets were improperly designed.

Lessons learned: Compression members must be checked against buckling. Confirmation of the strength of the existing concrete must be verified in designing epoxy anchors.

Incident No. 36
Elevated Structural Steel Platform

The structural engineer arbitrarily selected a single angle as a knee brace to support the platform.

Lessons learned: AISC has a special provision to determine the axial capacity of a single angle. If not followed, it could result in buckling and catastrophic failure.

Incident No. 56
Structural Steel Billboard

This incident occurred the day after the ironworkers completed the framing of a 60-foot-high billboard and its walkway. It resulted in three fatalities. The structure was meant to display advertisements along an interstate highway. The framing consisted of a hollow tube section cantilevered from a tall hollow tube column. At the end of the cantilever member, another long cantilever member was framed over a stub leg welded to the cantilevered beam. The failure occurred at the junction of the round stub leg and the long cantilevered beam.

An investigation revealed that the structural design was flawed in that it did not meet the AISC design specifications for hollow structural sections. The requirement to address chord wall plastification at the junction of the stub pipe and the cantilevered beam was not met. Also, the weld between the stub column and the cantilevered pipe was improperly performed.

Lessons learned: Differences between the design of rolled W shapes and HSS must be recognized, and the appropriate AISC provisions must be followed. Field testing of critical welds must be specified.

Incident No. 60
Scaffold Tower inside a Boiler

The structural engineer under-proportioned the steel beams supporting the 180-foot high scaffold tower inside a boiler.

Lessons learned: Light structural beams, when unbraced, are highly susceptible to lateral-torsional buckling. Proper design must evaluate the unbraced compression flange length.

Incident No. 61
Pedestrian Bridge

A long steel box girder spanning approximately 170 feet collapsed as the top concrete deck was being poured. At the time of the collapse, the concrete pour was almost halfway complete. Under the weight of the concrete, men and equipment, the girder twisted and collapsed. Though cross frames were provided between the girder top and bottom flanges, lateral bracing of the girder itself, when considered as a single longitudinal member, was not provided. As a result, the entire girder twisted in a torsional mode and fell.

Lessons learned: To prevent torsional buckling, long-span steel girders for bridges must have sufficient torsional rigidity in accordance with all applicable codes.

Incident No. 64
Custom Cantilever Finishing Platform

Loads considered in the design of the platform proved to be much lower than the actual loads placed on the platform.

Lessons learned: The engineer must consider realistic and verifiable loads while designing structural framing systems.

Incident No. 66
Parking Garage

A concrete structure under construction collapsed during placement of wet concrete on the 8th level, killing four construction employees and injuring scores of others. Five levels of exterior bays of the structure collapsed, with a complete separation from the exterior columns, while still being connected to the first row of interior columns. An investigation revealed that the exterior beam-column joints were improperly designed. Lack of sufficient reinforcement between the wide perimeter beams and the slender exterior leg contributed to the collapse.

Lessons learned: Exterior slender columns supporting wider beams must be avoided; otherwise, proper beam-column joint connections must be designed and detailed.

Incident No. 68
Gantry Crane

The structural design of the gantry crane was flawed in that it lacked a proper lateral-load-resisting system in one direction. During sudden braking of the crane, a large inertia force developed, and that force toppled the crane.

Lessons learned: Lateral-load-resisting systems must be provided in both directions. Lateral loads could arise from wind, seismic or inertia forces.

Incident No. 81
Stripping Platform

The structural design of the stripping platform was flawed. The temporary shoring frame of the platform was not properly supported, and the outer brace was incorrectly proportioned.

Lessons learned: Load paths and supports of platforms must be evaluated properly. Long compression members are highly susceptible to failure.

Incident No. 90
Post-tensioned Concrete Parking Garage

The structural design undersized the columns and under-proportioned the post-tensioned beam. A number of construction defects also existed.

Lessons learned: Exterior, as well as interior, slender columns supporting wider beams must be avoided; otherwise, proper beam-column joint connections must be designed and detailed.

Conclusion

To prevent structural collapses during construction due to design errors, the SER should:

- Design the entire structure, including components, using the latest applicable industry standards.
- Provide drawings with details or notes such as:
  - For concrete buildings, rebar development and splice lengths and placing details at congested areas; e.g., beam-column and beam-beam connections.
  - For steel buildings, steel yield strengths and welded connections, indicating length, size and type of welds.
  - For timber structures, lumber material with sizes and connection design requirements.
- Consider construction loads during design.
- Avoid exterior slender columns supporting wider perimeter beams.
- Ask for field testing of critical welds.
- Provide stability checks against lateral-torsional buckling.
- Evaluate vortex shedding for tall stacks.
- Check against seepage forces during sheet piling embedment design.
- Verify loads during final design.
- Caution the contractor to provide temporary bracing for unreinforced masonry walls, joists and roof trusses.
- Verify existing concrete strength, rather than making an assumption.
- Determine proper load paths during design.
- Follow AISC specification for single angle design.
- Provide proper bracing or rigid connections to resist lateral loads in both directions.■

Mohammad Ayub, P.E., S.E. is Director of Office of Engineering in the Directorate of Construction, Occupational Safety and Health Administration, US Department of Labor, Washington, DC.

Table 1: Summary Of Structural Failure Incidents

Type of Collapsed Structure & Incident number (X)	Most probable Cause/causes of the Collapse	Caused by construction or design flaws
Masonry Wall (1)	The CMU wall during construction was out of plumb, was not laterally braced, and was overloaded and overstressed.	Construction. 2 Killed, 16 Injured.
Structural Steel Building (2)	No temporary bracing. 32 mph wind pressure created overturning moments in excess of the flexural capacity of the column bases.	Construction. 1 Killed.
Steel Beams of a Steel Framed Building (3)	Incorrect steel erection procedure. ⅞" diameter high-strength bolts on a roof cantilever beam were overstressed.	Construction. 3 Killed.
Precast Concrete Beams (4)	Inadequate development length of #7 bottom rebars.	Design. 1 Killed, 1 Injured.
Precast Concrete Wall Panels (5)	Four precast concrete wall panels were not adequately braced to resist the lateral wind pressure.	Construction. 1 Killed, 2 Injured.
Soldier Beam and Lagging Cofferdam (6)	Spacer connecting the wale and soldier beam was not adequately designed to resist the expected earth pressures.	Design. 0 Injured.
Masonry Wall (7)	No temporary bracing. A gust of wind of 41 mph precipitated the failure.	Construction. 1 Killed, 8 Injured.
Steel Stack (8)	Vortex shedding was not considered in the design. It caused the fracture of the field fillet welds at the second splice from the top.	Design. 1 Killed, 6 Injured.
Concrete Shoring System (9)	Erected shoring system did not conform to the design drawings and was inadequate for carrying the imposed load during the placement of the concrete at the 4th level.	Construction. 3 Injured.
Roof Cable Structure (10)	Eccentrically applied tensioning force at the temporary jacking strands overloaded the gusset plate.	Construction. 1 Killed, 2 Injured.
Steel Erection Towers (11)	Inability to resist the forces imposed on the steel towers due to the movement of the west barge.	Construction. 1 Injured.
Structural Steel Framing (12)	Inadequate temporary connection of the members and placement of construction materials over the roof members.	Construction. 1 Killed, 1 Injured.
Structural Steel Column (13)	Lateral load applied at the top of the column overstressed the welds at the column base.	Construction. 2 Killed.
Overturning of Beams during the Cambering Operation (14)	Inadequate lateral torsional buckling strength of the beams, and a reduced modulus of elasticity during heat application.	Construction. 1 Killed.
Steel Sheeting Cofferdam (15)	Steel sheeting of the cofferdam did not have sufficient penetration to prevent soil failure.	Design. 2 Killed, 2 Injured.
Concrete Shoring System (16)	Formwork support system was underproportioned for both vertical and horizontal loads.	Design. 18 Injured.
Precast Concrete Single and Double Tees (17)	Unplanned additional dead load placed over the single and double tees caused the collapse.	Construction. 6 Injured.
Scaffold Frame (18)	Placement of brick at various tiers of the scaffold frame overstressed the frame.	Construction. 1 Killed, 4 Injured.
Movable Winch for Concrete Shaft Forms (19)	Winch unit was not fully assembled to its supporting beams with four hold-down rods, as required, before tensioning the wire rope of the winch.	Construction. 1 Killed, 4 Injured.

Reprinted and distributed with permission by STRUCTURE® magazine ■ December 2010 ■ www.STRUCTUREmag.org

350' high Radio Transmission Tower (20)	Extensive use of corroded tower sections salvaged from older towers. Snatch block at 22 feet above base created lateral loads overstressing the tower.	Construction. 1 Killed, 1 Injured.
Mast Climbing Work Platform (21)	The scaffold platform structure, as it was configured and erected in the field, was not designed for the loads imposed upon it.	Construction. 3 Killed, 2 Injured.
Steel Canopy Structure (22)	The canopy cantilever trusses and the back-up frames were not properly designed for the imposed loads.	Design. 1 Killed, 1 Injured.
Precast Concrete Parking Garage (23)	The shoring towers were neither diagonally braced nor plumbed. The shore frames were overloaded and overstressed.	Construction. 0 Injured.
Pre-engineered Metal Building (24)	Intermediate roof bent collapsed due to the lateral force created by a forklift to facilitate connections.	Construction. 2 Injured.
Structural Steel Joists (25)	The unbraced top chords of the joists caused the collapse. Though the joists were provided with 8 rows of diagonal bridging, the bridging lines were not anchored and were thus rendered ineffective.	Construction. 0 Injured.
1500-foot- high Antenna Tower (26)	Higher forces and moments were developed when the assembly of "track and gin pole" reached a higher elevation. Under this condition, the center of gravity of the assembly above the track underwent large rotations creating undue forces on the system as the bottom of the track was not anchored to the tower.	Construction. 3 Killed.
Masonry Foundation Wall (27)	The design of the concrete masonry wall was inadequate for both construction and permanent loads.	Design. 1 Killed.
Concrete Framed Building with Masonry Walls (Demolition) (28)	The shoring towers used for demolition had no diagonal bracings and were overloaded and overstressed.	Construction. 3 Killed
Wood Roof Light Gage Metal Building (29)	The timber roof trusses collapsed because they were not adequately braced.	Construction. 1 Killed, 1 Injured.
Escalator Truss (30)	Bolts connecting two escalator truss segments were overstressed. A number of bolts were only partially engaged.	Construction. 1 Killed.
Pre-engineered Metal Building (31)	Lack of adequate temporary bracing of the structural steel frames.	Construction. 1 Killed.
Steel Framed Parking Garage (32)	The steel erector did not comply with the sequence of erection and procedures recommended by his consultant. The steel erector failed to provide temporary bracing and guy wires.	Construction. 3 Killed.
Bridge Road (33)	The north steel brackets on the piers of the center span of the bridge were improperly designed to support the intended loads.	Design. 1 Killed, 1 Injured.
Grain Elevator (34)	The steel erector did not follow proper erection procedures from the manufacturer. Guy cables or temporary supports were not provided during the erection of the elevator.	Construction. 1 Killed.
1,889' high TV Antenna Tower (35)	A diagonal member of the tower was removed by the contractor without using any come-a-longs. This resulted in overstressing of the tower members and buckling of tower legs.	Construction. 3 Killed.
Elevated Structural Steel Platform (36)	The platform failure was caused by the inadequate temporary knee bracing to support the loads on the platform.	Design. 1 Killed, 4 Injured.
Reinforced Concrete Building (Demolition) (37)	Engineering survey was not performed to determine the height of the elevator shaft and the weight of two steel drums over the elevator. The elevator shaft fell onto the excavator cab and killed the operator.	Construction. 1 Killed.
Steel Bridge Girder (38)	The C-clamps used to temporarily hold the cross frames to the steel girder were not adequate. The 4x6 temporary strut and come-a-long wire to support the collapsed girder were also inadequate.	Construction 2 Injured.

Reprinted and distributed with permission by STRUCTURE® magazine ■ December 2010 ■ www.STRUCTUREmag.org

Scaffold Tower (39)	The scaffold tower was erected without any diagonal bracing at some locations. Also, the K-braces and X-braces were removed at a few locations. In addition, the lateral bracings were temporarily removed at certain levels.	Design and Construction. 1 Killed.
Guardrail System (40)	Guardrail system was not properly designed and installed to withstand a 200-pound load.	Construction. 1 Killed.
Reinforcing Steel Cage (41)	The erector did not provide adequate lateral support to the 44-foot-high reinforcing steel cage.	Construction. 1 Killed.
Structural Steel Framed Building (42)	The contractor removed the temporary bracings of the building frame prematurely. The contractor did not follow the written instructions from the design engineers.	Construction. 2 Killed, 2 Injured.
Floor Beams of a Structural Steel Framed Building (43)	Due to the lack of welding between the metal deck and steel beams, one of the interior steel beams was overstressed and collapsed during the concrete pour operation.	Construction. 4 Injured.
Scaffold Towers (44)	Lateral braces were not provided between the putlogs. Scaffold towers were erected with components from different manufacturers. The contractor did not perform any structural analyses.	Construction. 3 Injured.
Reinforced Concrete Building (Demolition) (45)	The contractor imposed loads on the fourth floor in excess of its ultimate capacity by placing an excavator over 2-foot-deep debris.	Construction. 1 Injured.
Wood Framed Apartment Building (46)	Washer plates on top of the sill plate were not provided. The exterior sheathing was not nailed to the bottom sill plate. The building collapsed during a windstorm.	Construction. 1 Killed, 4 Injured.
Bridge Paint Containment Structure (47)	The fabricator of the containment structure did not install the proper size U-bolts as per contract documents, and did not perform the welding using certified welders.	Construction. 1 Killed, 2 Injured.
Longspan Steel Roof Joists (48)	The bridging lines of four roof joists were not properly anchored to transfer the lateral loads.	Construction. 1 Killed.
Steel Roof Trusses (49)	The steel erector did not provide adequate temporary bracing to the roof trusses. The bridging lines of trusses were not continuous and were not properly anchored.	Construction. 0 Injured.
Masonry Wall and Wood Floor Framing Apartments (50)	The mason contractor overloaded the second floor beyond its capacity by placing masonry blocks.	Construction. 5 Injured.
Steel Sheeting Cofferdam (51)	The contractor arbitrarily eliminated and changed sizes of a number of main structural members without informing the engineer of record. It resulted in a deficient and unstable structure.	Construction. 0 Injured.
Mast Climbing Platform (52)	The overloading of the scaffold platform caused the collapse. The four sections of the platform were loaded well in excess of their safe capacities.	Construction. 3 Injured.
Scaffold Towers (53)	The scaffold towers were overloaded and overstressed beyond their ultimate capacity.	Construction. 5 Killed, 10 Injured.
Wood Roof Trusses (54)	Inadequate lateral and diagonal bracings were provided at the top chords of the failed trusses. There were no lateral and diagonal bracings at the bottom chord.	Construction. 5 Injured.
Steel Joists Floor Framing (55)	The steel joists were not laterally braced and failed under lateral torsional buckling.	Construction. 1 Killed, 4 Injured.
Structural Steel Framed Billboard (56)	The structural design of the steel framing did not follow the requirements of AISC's *Hollow Structural Sections*. The weld at the junction of the stub pipe and cantilever beam was not properly performed.	Design & Construction. 3 Killed.
Tilt-up Precast Concrete Wall Panel (57)	The failure occurred because the contractor prematurely removed the temporary braces of the tilt-up wall panel before permanent connections at the top and bottom of the panel were made.	Construction. 3 Killed.

Reinforced Masonry Wall (58)	The masonry contractor did not provide adequate bracings to prevent collapse of the wall. The bracings were deficient because they were neither properly proportioned nor connected.	Construction. 1 Killed.
1,965' high TV Antenna Tower (59)	The contractor removed the bolts connecting the diagonals, a horizontal strut and redundant members resulting in a significant reduction of the load-carrying capacity of the tower legs.	Construction. 2 Killed, 3 Injured.
Scaffold Inside a Boiler (60)	The selection and design of the base beams supporting the scaffold were flawed. The beams were significantly undersized and not capable of supporting the intended load.	Design. 2 Injured.
Pedestrian Bridge (61)	The stability of the bridge girder during construction and placement of wet concrete was not considered by the structural engineer of record. Adequate top lateral bracings were not provided.	Design. 1 Killed, 9 Injured.
Light Gage Metal Framed Building (62)	The incident occurred because the contractor failed to provide proper lateral bracings to the studs by means of straps.	Construction. 0 Injured.
Wood Roof Trusses (63)	The truss erector did not provide temporary bracings during erection, as per the industry standard.	Construction. 1 Killed.
Custom Cantilever Finishing Platform (64)	The scaffold was overloaded beyond its rated load. In addition, the imposed loads were not uniformly distributed over the platform; instead, they were concentrated at one area.	Design. 1 Killed. 3 Injured.
1,000foot-high TV Antenna Tower (65)	The contractor placed a hoist (winch) at a location that produced eccentricity between the load and the center of gravity of the tower. The contractor did not perform a structural evaluation of the tower to determine the maximum tension in the load line that could safely be applied.	Construction. 3 Killed.
Reinforced Concrete Parking Garage (66)	The structural design of the slab/beam-column joints was flawed. The concrete contractor did not provide the required embedment length for the welded wire mesh at the intersection of the exterior columns and the exterior edge of the beam. The concrete contractor failed to detail, fabricate and place bottom reinforcing steel.	Design and Construction. 4 Killed, 21 Injured.
Steel Bridge Girder (Demolition) (67)	Inadequate lateral torsional buckling strength of girder overstressed the girder and led to collapse.	Construction 1 Killed.
Gantry Crane (68)	The structural design was flawed as it did not provide a proper lateral load-resisting system in the north-south direction. A large inertia force that occurred when the crane suddenly stopped caused the crane to fail.	Design. 1 Killed.
Balcony of an Apartment Building (69)	Improper embedment of an overhang balcony beam inside CMU wall caused the balcony to collapse under wet concrete.	Construction. 1 Killed, 2 Injured.
Townhouse Building (70)	Lack of an adequate number of single post shores under the floor to be cast. Improper placement of tunnel formworks, i.e., leveling jacks were not turned down to the floor slab to transfer the load.	Construction. 2 Killed, 3 Injured.
Scaffold Towers (71)	The scaffold was overloaded. Scaffold legs were not properly braced, which significantly reduced its compressive capacity.	Construction 4 Injured.
Precast Concrete Parking Garage (72)	Lack of horizontal bracing for the double tee during the jacking process. Lack of lateral bracing for the exterior columns.	Construction 1 Killed.
Steel Rebar Cage (73)	Inadequate lateral support for the rebar cage. Lack of use of standard procedure for the installation of rebar cage.	Construction 1 Killed, 3 Injured.
Steel Framed Craneway (Demolition) (74)	The flamed cuts made to the steel frames rendered them unstable and eventually resulted in an unplanned collapse.	Construction 2 Killed, 4 Injured.
Wood Roof Trusses (75)	Inadequate lateral and diagonal bracings for the wood roof truss.	Construction 5 Injured.
Concrete Roof with Masonry Wall Building (Demolition) (76)	Lack of temporary bracing and supports during the demolition of the central portion of the building.	Construction 0 Injured.

Concrete Formwork (77)	Aluminum shoring legs were overloaded beyond their rated loads. Beams were not centered on U-head. Some joists/stringers were oriented to weak axis.	Construction 10 Injured.
Steel Roof Trusses (78)	The steel erector did not follow the generally accepted standard practice to provide stability against lateral-torsional buckling to the girder truss during erection.	Construction 2 Killed, 2 Injured.
Concrete Formwork (79)	The collapse was triggered when the defective screw jack at one of the shoring legs failed under an eccentric load, caused by the off-centered stringer inside the U-head.	Construction 0 Injured.
Wood Roof Trusses (80)	The truss erector did not provide proper bracings to the long span trusses, in violation of the standard industry practice.	Construction 1 Killed.
Stripping Platform (81)	The structural design of the tunnel forms was flawed. The false frame was not appropriately supported and the outer brace was not correctly proportioned.	Design 2 Killed.
Concrete Formwork (82)	The concrete subcontractor did not erect the formwork of table No. 15 as per the approved formwork drawings. Other contributing factors were improper reshores, damaged truss members, fewer and out of plumb screw jacks, and improper mudsills.	Construction 3 Killed, 2 Injured.
Bridge Truss (Demolition) (83)	The demolition of the last two spans of the bridge was carried out in such a way that the primary structural member was overstressed beyond its ultimate strength.	Construction 1 Killed, 1 Injured.
Concrete Formwork (84)	The contractor proceeded to cast concrete on the roof level without having any formwork design drawings. The contractor arbitrarily determined the roof formwork framing without any evaluation. The contractor used damaged aluminum beams in the roof formwork.	Construction 17 Injured.
Precast Concrete Parking Garage (85)	A number of steel columns of the parking garage were erected out of plumb, beyond the permitted tolerances. The failed precast plank was installed with only 1 inch of bearing.	Construction 1 Killed.
Masonry Wall (86)	Lack of lateral bracing. Wind gusts caused the collapse of the improperly constructed and unbraced block wall.	Construction 0 Injured.
Concrete Formwork (87)	Early stripping of the beam soffit, reduced concrete cover below the embed, and overloading of an embed during leveling.	Construction 1 Killed, 1 Injured.
Concrete Column Form (88)	The contractor used a concrete mix which contained admixtures that yielded a lateral pressure on the formwork greater than its capacity. Also, rate of pouring was too fast.	Construction 2 Killed.
Wood Roof Trusses (89)	Lack of temporary bracing and overloading of roof trusses by placing a stack of materials during erection.	Construction 2 Injured.
Post-tensioned Concrete Parking Garage (90)	Flawed structural design for beams and columns, and lack of reshores below the third level.	Design and Construction 1 Killed.
Tower Crane (91)	Use of polyester and deteriorated slings to suspend the collar, and improper rigging of the collar. Also, slings were not protected against sharp edges for cuts and abrasions.	Construction 7 Killed.
Slab on Grade (Demolition) (92)	Lack of underpinning under the CMU wall.	Construction 1 Killed.
Suspended Scaffold (93)	The scaffold was overloaded. Due to the traffic vibrations of the bridge deck, the dynamic effect of the suspended swivel bolt significantly contributed to the failure of the supporting tongue plate.	Construction 3 Killed.
Concrete Ramp (94)	Lack of support of the precast joists at intermediate points when the slab of the ramp was being poured. The contractor used the 4x4 wooden shores with Ellis Jacks in lieu of required steel shores as per the formwork drawing.	Construction 1 Killed, 1 Injured.
Mobile Crane (95)	The crane manufacturer did not design the boom stops to prevent the boom from falling backwards.	Design. 4 Killed, 6 Injured.
Gantry Crane Leg (96)	During dismantling of the crane leg, excessive forces were applied at the toe of the leg, which severed its upper supports. As a result, the leg came crashing to the ground.	Construction 1 Killed, 3 Injured.

Foundations

- Shallow
  - Mat
  - Floating
  - Strip Footings
  - Column Footings
- Deep
  - Piles
  - Piers

Photo courtesy of John Gambatese

Reinforced Concrete

NOTES

Foundations are typically classified as shallow or deep. Shallow foundations use column footings, strip footings, mat foundations, or floating foundations. The design of footings, including the amount of reinforcement and the type of footing used, depends largely on the type of structure being erected as well as the soil conditions.

Deep foundations use piles or piers, which are slender elements (columns) buried deep in the soil. Deep foundations are used where a typical shallow foundation will not hold up in the soil conditions, such as sand or some other unstable soil.

Piers are most related to reinforced concrete construction because they are usually site-cast in predrilled holes. Once the hole is drilled to specification, a prefabricated rebar cage is placed in the hole and then the pier concrete can be poured.

SOURCE

Photo courtesy of John Gambatese

Reinforcement

- Concrete is about 90% weaker in tension than it is in compression

- Steel has high tensile strength, has the same thermal expansion as concrete, and bonds well with concrete

Photo courtesy of John Gambatese

NIOSH [2010]. Reducing work-related musculoskeletal disorders among rodbusters
www.cdc.gov/niosh/docs/wp-solutions/2010-103.

Reinforced Concrete

NOTES

Reinforcement is needed in structural concrete applications because concrete is about 90% weaker in tension than it is in compression. Reinforcing steel is added to resist tension, while the concrete is counted on to resist compression. Because steel has high tensile strength, has the same thermal expansion as concrete, and bonds well with concrete, it is a great complementary material to make concrete more versatile. Rebar can be made with billet steel, rail steel, or axle steel, and it almost always is a deformed bar in order to create a mechanical bond with the concrete once the concrete cures. Some rebar is epoxy coated for corrosive environments such as places where deicers are commonly used (roadways and bridge decks).

Another form of reinforcement commonly seen in buildings is welded wire reinforcement (WWR). This reinforcement is used for slabs when the need for reinforcement is marginal. WWR is also sometimes called mat reinforcement. In some cases the reinforcement may be placed within sheaths and stressed after the concrete is cured. This is called pre-stressed or post-tensioned concrete. In this case, the reinforcement is high-strength cables draped along the member to best resist the internal member forces.

NIOSH evaluated reinforcing ironworkers' (rodbusters) exposures to risk factors for developing low-back and hand disorders when tying together reinforcing steel bars (rebar) on a freeway bridge. Rodbusters used three techniques to tie rebar together—a pliers and a tie wire wheel, a battery operated power tier (PT), and a PT with an extension handle (PTE). NIOSH found that using the PT and PTE reduced the rodbusters' exposures to risk factors for work-related low-back and hand-wrist disorders. In addition, power tying was twice as fast as pliers tying.

SOURCES

Photo courtesy of John Gambatese

NIOSH [2010]. Reducing work-related musculoskeletal disorders among rodbusters. Cincinnati, OH: U.S. Department of Health and Human Services, Centers for Disease Control and Prevention, National Institute for Occupational Safety and Health, DHHS (NIOSH) Publication No. 2010–103 [www.cdc.gov/niosh/docs/wp-solutions/2010-103/].

WORKPLACE SOLUTIONS

From the National Institute for Occupational Safety and Health

Reducing Work-Related Musculoskeletal Disorders among Rodbusters

Summary

NIOSH evaluated reinforcing ironworkers' (rodbusters) exposures to risk factors for developing low-back and hand disorders when tying together reinforcing steel bars (rebar) on a freeway bridge. Rodbusters used three techniques to tie rebar together—a pliers and a tie wire wheel, a battery operated power tier (PT), and a PT with an extension handle (PTE). NIOSH found that using the PT and PTE reduced the rodbusters' exposures to risk factors for work-related low-back and hand-wrist disorders. In addition, power tying was twice as fast as than pliers tying.

Description of Exposure

Reinforcing ironworkers have reported high prevalence rates for work-related musculoskeletal disorders (WMSD) symptoms affecting the low-back (80.2%) and wrists/hands (48.4%) [Cook et al 1996]. Boston-area rodbusters reported high prevalence rates for self-reported symptoms of the low-back (52.2%) and hands/wrists/fingers (47.8%) and high prevalence rates of doctor-diagnosed WMSDs, including ruptured spinal discs (14%) and carpal tunnel syndrome (16%) [Forde et al. 2005].

Traditionally, *pliers and a tie wire wheel* have been used to pull, wrap, twist, and cut the 'tie' wire around two or more concrete reinforcing bars. This requires using both hands and making rapid and repetitive hand, wrist, and forearm movements while gripping the pliers. In recent years, *power tiers* have become available. The PT is a battery-powered and trigger-operated wire tier that automatically wraps, cuts, and ties the wire around the rebar. Tying rebar at ground level using either the pliers or the PT requires working in a stooped posture. A height-adjustable extension handle (PTE) is commercially available for one type of handheld PT enabling the worker to tie the rebar while standing.

Evaluation

A concrete reinforcing contractor requested that NIOSH evaluate workers' exposures to WMSD risk factors during rebar tying on a freeway bridge deck construction project that required making more than 2 million "ties." The contractor's workers used both pliers and PTs to tie rebar. NIOSH introduced the PTE as a third technique to be investigated in the study. Although rodbusters perform other job activities that require "maximum muscle force to lift, push, pull, or carry objects" [ONET 2008], NIOSH analyzed only rebar tying during this study because of the nature of the request and time constraints.

The three rebar tying methods were studied with relation to (1) hand, wrist, and forearm position and movements and (2) trunk (or back) position.

Results

Hand/Wrist

- *Pliers* tying involved the most hand, wrist, and forearm motions and the highest risk for developing a WMSD of the hand-wrist (see Figure 1).

- *PT* and *PTE* tying involved fewer hand, wrist, and forearm motions and less risk for developing a WMSD of the hand-wrist.

Figure 1. Rebar tying using pliers. Note the bent posture and awkward hand position.

- Workers rated hand-wrist effort highest for *pliers* and the *PTE*, and least for *PT* tying.

Low-back

- *Pliers* tying at ground level involved the most risk for low-back problems.
- *PT* tying at ground-level allowed workers to support the weight of their trunk with one hand.
- *PTE* tying could be done standing-up with the least strain on the low-back.
- Workers said they had the most low-back discomfort using *pliers* and the least using the *PTE* and the *PT*.

NIOSH found that the frequency and duration of the hand and wrist motions are associated with increased risk of a hand-wrist WMSDs [NIOSH 2005]. The similar effort ratings for tying with pliers and the PTE conflict with the results of other studies in which workers having experience using the extension handle reported much less effort [Vi 2003]. Workers did not have time to use the extension handle before this study began. Consequently, workers were observed holding the PTE far from the body, which would increase the stress on the shoulder, elbow, and wrist.

Workers reported less low-back effort using the PT than the pliers, although both required frequent and prolonged stooping (see Figure 2). During the study, NIOSH observed all workers using the free hand/arm—the one not holding the PT—to support their body when stooping. This posture likely reduced the stress to the low back and the report of less low-back effort, which is consistent with other reports [Gallagher et al 1988; Ferguson 2002; Kingma 2004].

Tying rebar using the PT and PTE was faster than using the pliers. Workers were able to complete twice as many ties during the study period with the PT as compared with the pliers. Vi [2005] reported that PTE tying by experienced workers was twice as fast as pliers tying. Contractors and workers have reported difficulties with power tier use that can affect actual productivity levels, including tool malfunction, wire jams, and short battery life [ORISE 2007]. Power tiers can make one type of tie and are not appropriate for all applications.

Recommendations

Contractors and workers should take the following steps to reduce the risk of developing MSDs when tying rebar on freeway bridge decks and other construction

Figure 2. Rebar tying using a MAX–USA RB–392 power tool. Note the bent posture.

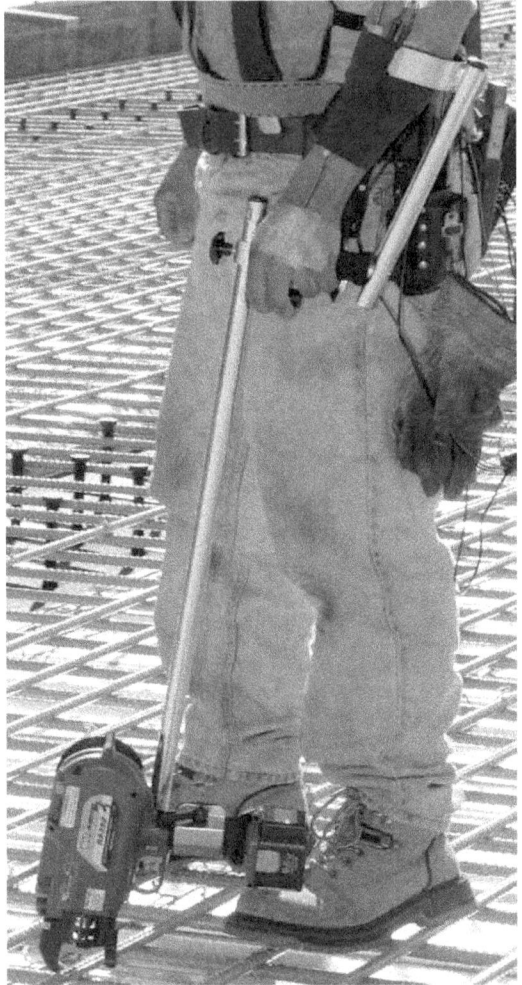

Figure 3. Rebar tying using a MAX–USA RB–392 power tool with adjustable extension.

projects requiring frequent and prolonged rebar tying. [Albers and Hudock 2007; NIOSH 2005]:

- Use PTs instead of pliers to reduce harmful hand-wrist movements.

- Use a PTE when tying ground level rebar.

- When using a PTE, hold it close to the body to avoid unnecessary stress and strain on the wrist, arm, and shoulder (see Figure 3).

- When using a PTE extension, adjust the height of the hand-grip so that it can be firmly held with your arm hanging relaxed to minimize stress on the upper extremities and low back.

- Report low-back or upper-limb aches, stiffness, or pain that may be due to your work to your health care provider.

Acknowledgments

This document was prepared by Jim Albers, MPH, CIH and Stephen D. Hudock, Ph.D., CSP, Division of Applied Research and Technology, National Institute for Occupational Safety and Health.

References

Albers JT, Hudock SD [2007]. Biomechanical assessment of three rebar tying techniques. Int J Occup Safety Ergonomics 13(3):227–237.

Cook TM, Rosecrance JC, Zimmerman CL [1996]. The University of Iowa construction survey. Washington, DC: Center to Protect Workers' Rights, Report No. E1–96.

Forde MS, Punnett L, Wegman DH [2005]. Prevalence of musculoskeletal disorders in union ironworkers. J Occup Environ Hygiene 2:203–212.

Ferguson FA, Gaudes-MacClaren LL, Marras WS, Waters, TA, Davis KG [2002]. Spinal loading when lifting from industrial storage bins. Ergonomics 45(6):399–414.

Forde M, Buchholz B [2004]. Task content and physical ergonomic risk factors in construction ironwork. Int J Ind Ergon 34:319–333.

Gallagher S, Marras WS, Bobick TG. [1988] Lifting in stooped and kneeling postures: effects on lifting capacity, metabolic costs, and electromyography of eight trunk muscles. Int J Ind Ergonomics 3(1):65–76.

Kingma I, van Dieen JH [2004]. Lifting over an obstacle: effects of one-handed lifting and hand support on trunk kinematics and low back loading. J Biomechanics 37:249–255.

NIOSH [2005]. NIOSH Health Hazard Evaluation Report: Genesis Steel Services, Inc., Baltimore, MD. Cincinnati, OH: U.S. Department of Health and Human Services, Centers for Disease Control and Prevention, National Institute for Occupational Safety and Health, NIOSH HHE Report 2003–0146–2976 [www.cdc.gov/niosh/hhe/reports/pdfs/2003-0146-2976.pdf].

ONET [2008]. Reinforcing iron and rebar workers (47–2171.00). Occupational Information Network (O*NET) Online http://online.onetcenter.org/link/summary/47-2171.00. Accessed April 17, 2008.

ORISE [2007]. Message testing and formative research on perceived stakeholder barriers and incentives for the adoption and diffusion of concrete reinforcement and placement interventions: focus group report. Oak Ridge, TN: Oak Ridge Institute for Science and Education.

Vi P [2003]. Reducing risk of musculoskeletal disorders through the use of rebar-tying machines. Appl Occup Environ Hygiene 18(9):649–654.

DEPARTMENT OF HEALTH AND HUMAN SERVICES
Centers for Disease Control and Prevention
National Institute for Occupational Safety and Health
4676 Columbia Parkway
Cincinnati, OH 45226–1998

Official Business
Penalty for Private Use $300

For More Information

More information about ergonomics in construction is available from the NIOSH publication

NIOSH [2007]. Simple solutions: ergonomics for construction workers. By Albers J and Estill C. Cincinnati, OH: U.S. Department of Health and Human Services, Centers for Disease Control and Prevention, National Institute for Occupational Safety and Health, DHHS (NIOSH) Publication No. 2007-122 [http://www.cdc.gov/niosh/docs/2007-122/].

For general information about construction safety and health topics, visit the NIOSH Construction Topic Page at http://www.cdc.gov/niosh/topics/construction.

For general information about musculoskeletal disorders and carpal tunnel syndrome, visit this NIOSH Topic Page: http://www.cdc.gov/niosh/topics/ergonomics/

To receive information about other occupational safety and health topics, contact NIOSH at

Telephone: 1–800–CDC–INFO (1–800–232–4636)
TTY: 1–888–232–6348 • E-mail: cdcinfo@cdc.gov

or visit the NIOSH Web site at http://www.cdc.gov/niosh

For a monthly update on news at NIOSH, subscribe to *NIOSH eNews* by visiting http://www.cdc.gov/niosh/eNews.

Mention of any company or product does not constitute endorsement by NIOSH. In addition, citations to Web sites external to NIOSH do not constitute NIOSH endorsement of the sponsoring organizations or their programs or products. Furthermore, NIOSH is not responsible for the content of these Web sites.

This document is in the public domain and may be freely copied or reprinted. NIOSH encourages all readers of the *Workplace Solutions* to make them available to all interested employers and workers.

As part of the Centers for Disease Control and Prevention, NIOSH is the Federal agency responsible for conducting research and making recommendations to prevent work-related illness and injuries. All *Workplace Solutions* are based on research studies that show how worker exposures to hazardous agents or activities can be significantly reduced.

Reducing Work-Related Musculoskeletal Disorders among Rodbusters

DHHS (NIOSH) Publication No. 2010–103

SAFER • HEALTHIER • PEOPLE™

November 2009

 **Slabs**

- On-Grade
  - Isolated
  - Stiffened
- Elevated Slabs
  - Beam-supported
  - Beamless
  - Extensive formwork

Photo courtesy of John Gambatese

Reinforced Concrete

NOTES

There are two types of concrete slabs: ground supported (slab-on-grade) and elevated slabs. For slabs-on-grade, there are also two types: isolated concrete slabs and stiffened concrete slabs. Isolated concrete slabs are separated from the foundation by isolation joints, whereas stiffened concrete slabs are used as a foundation. Typically, a vapor retarder is placed underneath the slab-on-grade, which prevents water from seeping up through the slab into the building's interior. Slabs-on-grade typically do not require an immense amount of reinforcement because they are supported continuously by the soil. However, reinforcement is still used to reduce cracking and allow for further control of joint spacing. Elevated slabs consist of beam-supported floors and beamless slabs. Beam-supported concrete slabs have beams and girders for support. Elevated slabs require much more reinforcement than slabs-on-grade because of the amount of tension the slabs experience. In order to pour the slab, a significant amount of formwork is needed, as well as shoring to hold the elevated slab in the air. Once the slab has cured to the proper strength, the shores can be removed and can be replaced with reshores if necessary.

SOURCE

Photo courtesy of John Gambatese

Beams and Girders

- For simple spans:

  - Tension in bottom of beam

  - Compression in top of beam

- Precast elements tied into buildings with hooks, lap splices, or couplers

Photo courtesy of Thinkstock

Reinforced Concrete

NOTES

Cast-in-place concrete beams and girders are formed and poured the same way as elevated slabs. Significant shoring is required, as well as reinforcement. Because the bottom of the beam experiences the most tension for simply supported members, this is where most of the rebar is located. It is not unusual for a concrete beam to have larger reinforcement at the bottom and smaller reinforcement at the top in order to save on steel costs. Rebar is held together by stirrups, which are wrapped around the rebar and hanger bars. Precast beams and girders are tied into the building by means of hooks, lap splices, or couplers.

SOURCE

Photo courtesy of Thinkstock

Columns

- Typically designed for compression, but must be able to resist bending

- Longitudinal rebar runs vertically and is held in place by ties

    – Longitudinal bars are typically about 4% of the gross column area; ties are usually #3 or #4 bars

Photo courtesy of John Gambatese

Reinforced Concrete

NOTES

Columns experience both compressive forces and bending. Longitudinal reinforcing bars run vertically and are held in place by ties. The ties prevent buckling of the longitudinal reinforcing bars and resist shear forces during bending. Longitudinal bars are typically around 4% of the gross column area. Ties are #3 or #4 bars. Forms for columns are similar to those of walls; however, they often have a hinge so that they can be unhooked and removed from the wall and reused easier. Round columns sometimes have spiral reinforcement and are formed with steel plate or waterproof fiber board.

SOURCE

Photo courtesy of John Gambatese

 **Walls**

- Concrete walls resist compression forces.

- Walls are reinforced with a mesh of vertical and horizontal rebar in a layer on each wall face.

- Formwork and form ties are used to ensure proper wall thickness.

Photo courtesy of John Gambatese

Reinforced Concrete

NOTES

Concrete walls in buildings are often used as retaining walls, load-bearing walls, basement walls, and shear walls. Typically walls use a mesh of reinforcement on each face, which consists of vertical and horizontal rebar in a layer. Thicker walls usually require multiple layers of rebar. The wall is formed with standard plywood formwork around the reinforcing layers. Commonly, form ties are used to hold each side of the form together to keep them from blowing out, as well as to ensure proper wall thickness. Additionally, walers and stiffbacks are used to brace the plywood forms. While the concrete is poured, it has to be vibrated between lifts to ensure proper settlement and to evacuate air bubbles.

Tilt-up walls consist of site-cast concrete panels that are then lifted into place. This method of construction has typically been used for industrial buildings but has become more popular because of the speed and cost-effectiveness. The walls are formed on the ground, similar to slabs, with lift points tied to the reinforcement. Cranes can then lift the slab to its final placement, where it is tied to the rest of the building.

SOURCE

Photo courtesy of John Gambatese

Pre-stressed Concrete

- Pre-tensioning
  - Cast over tensioned strands
- Post-tensioning
  - Cast over sleeves and tendons
  - Tendons are tensioned after slab cures

Photo courtesy of John Gambatese

Reinforced Concrete

NOTES

Prestressing concrete can be accomplished in two ways: pretensioning and post-tensioning. This is an alternative method to traditional reinforcing because it introduces compression into the members where tension is expected to be created by the loads. Pretensioned members are created by casting the concrete member over strands that are tensioned with hydraulic jacks, and as the concrete cures, a bond is formed between the strands and the concrete. This creates an upward curvature in the concrete member, which counteracts the applied downward load.

Post-tensioning is similar to pretensioning, but the strands are tensioned after the concrete has been placed. Strands are placed in sleeves, and when the concrete reaches sufficient strength, the tendons can be tensioned and anchored in that position by mechanical wedges. In post-tensioned members, the strands are free to move within the sleeve rather than being bonded to the concrete, which allows them to be tensioned without applying tension to the concrete.

SOURCE

Photo courtesy of John Gambatese

Precast Concrete

- Cast off-site and transported

- Reduces formwork and allows for curing in a controlled environment

- Increased transportation and hoisting costs

Photo courtesy of Thinkstock

Reinforced Concrete

NOTES

Precast concrete elements are those that are cast off-site and then transported to their final place. Precast concrete has the advantage of being cast in a controlled environment, which allows for more intricate shapes and often quicker production because of the reduction of formwork and shoring, as well as steam-curing. The disadvantage is additional transportation costs and hoisting requirements.

SOURCE

Photo courtesy of Thinkstock

Retaining Walls

- Walls made to withstand lateral earth pressure exerted by sloped soils
- Types
    - Gravity
    - Semi-gravity
    - Cantilever
    - Counterfort

Photo courtesy of Thinkstock

Reinforced Concrete

NOTES

Reinforced concrete is often used for retaining walls because of its ability to withstand the lateral earth pressure that sloped soils exert. Concrete retaining walls come in a number of types, including gravity retaining walls, semigravity retaining walls, cantilever retaining walls, and counterfort retaining walls.

SOURCE

Photo courtesy of Thinkstock

Reinforced Concrete Design Process

- Initial Design
- Shop Drawings
- Shop Drawing Submittal
- Shop Drawing Review
- Fabrication

Photo courtesy of John Gambatese

Reinforced Concrete

NOTES

Initial design is typically carried out by the architect and engineering design consultants. They determine the overall plan for the building, including the structure type. Their designs are very thorough, but these are not final. When the construction documents are issued, the concrete subcontractor bids the documents. Once the project is awarded, changes to the design may be made.

During the submittal process, contractors are required to send the architect product data about everything they are going to install on the building. For some specialties, these data are more involved than for others. Shop-drawing submittal is required of many contractors, including concrete contractors, with regard to their reinforcement. Fabricators are required to generate shop drawings that show exactly how they plan to build each concrete element, including rebar placement. Sometimes the fabricator employs its own engineer to design the reinforcement.

Once the shop drawings are completed, the fabricator sends the shop drawings and submittals to the general contractor for review of compliance with the specifications. If the contractor thinks they are within specification, then they are sent to the architect for review. The architect and the structural engineering consultant review the proposed design and either accept it or reject it and ask for a resubmittal. When the accepted submittal makes it back, the fabricator may begin to procure the materials and start fabrication and construction.

SOURCE

Photo courtesy of John Gambatese

Concrete Construction Activities

- Layout
- Rebar Installation
- Formwork
- Concrete placement
  - Batching
  - Mixing
  - Transporting
  - Placing
- Vibration
- Curing
- Form stripping

Photo courtesy of Walter Heckel

Reinforced Concrete

NOTES

The first step of setting up for a concrete pour is the layout. Formwork and rebar are built or placed in the desired location. Sometimes this process has to be done simultaneously because hook lengths of rebar extending into a wall (or another interaction between the reinforcements in two members) would block placement. Once rebar is in place, it is tied with wire to keep it from moving while the concrete is poured.

Concrete placement involves four steps: batching, mixing, transporting, and placing. Batching involves combining the correct volumes of the ingredients for a batch of concrete. In order to produce quality concrete, the batching must be consistent. Batching may be done by weight to ensure accuracy, but typically it is done volumetrically on site. Mixing concrete is an important process to ensure uniformity. This process, in addition to batching, ensures the proper slump is achieved, which gives the concrete its desired workability and strength. Transporting concrete is typically done by trucks. Each standard concrete truck carries 8 to 12 cubic yards of concrete. The method used for placing concrete depends on the amount of concrete being placed. Pump trucks are used for large pours, whereas crane buckets are used on sites with restricted access. Some pours, such as ground slabs, can be made directly from the concrete truck. Smaller jobs are carried out with a wheelbarrow and a shovel.

When concrete is being placed, it is typically vibrated to increase consolidation. This makes the concrete within a form system uniform and better suited for structural application. Additionally, the vibrating keeps the aggregate from getting stuck in the reinforcement.

Concrete strengthens with age. As concrete cures, it gains over half its strength in about a week and over 90% of its strength in 27 days. Improper curing can result in increased cracking and reduced strength.

Once concrete has reached sufficient strength, the forms can be stripped. Special care needs to be taken so that the forms are not removed early, because this can cause sagging and cracking. If the formwork involves shores for an elevated member, reshoring may be necessary to keep the structure from collapsing. Proper design of the size and spacing of reshores is necessary to ensure the reshore method is structurally adequate and cost-effective.

SOURCE

Photo courtesy of Walter Heckel

Hazards Associated with Reinforced Concrete Construction

Concrete Construction Hazards

- Tripping
- Muscle strain caused by repeated lifting
- Structural collapse
- Falling materials
- Manipulation and erection of reinforcing steel and formwork
- Silicosis

Photo courtesy of John Gambatese

Silicosis is caused from inhaling silica dust during concrete mixing, grinding, polishing or cutting
www.cdc.gov/niosh/docs/wp-solutions/2009-115/; www.cdc.gov/niosh/docs/wp-solutions/2008-127

CDC NIOSH   Reinforced Concrete

NOTES

Tripping is a significant risk associated with reinforced concrete and is largely due to the rebar placement. A battery-operated power tier with an extension handle can reduce hand/wrist and low-back injuries.

Rebar is often put in a lay-down area on site and then transferred to the final location. Rebar spacing in slab construction is a common tripping hazard because workers have to balance as they navigate across the reinforcement. The added danger with tripping around rebar is the potential for a puncture wound from any vertical steel around the site.

Collapse hazards are common in concrete construction because of the sheer weight of the material. Typically, collapse involves formwork. If formwork is improperly designed, it can blow out because of the lateral pressure from the weight of the wet concrete. Additionally, when collapse results from stripping the shoring too early on elevated slabs, there is substantially more danger because often people are working above or below the curing slab. Live loads experienced during construction as heavy equipment moves over the structure can also cause the structure to collapse.

Muscle strain caused by repeated lifting is a common injury because of the repetitive nature of the work. Workers have to constantly maneuver the rebar and repeatedly bend over to tie the rebar in place, which causes back strain. The steel can also be extremely heavy when the reinforcement is prefabricated, such as a spiral cage, and has to be lifted into place. When a crane is used to place large sections of formwork or reinforcement, a crushing hazard is created.

Falling materials, such as masonry units from an elevated installation, create a significant hazard on the job site. This danger is usually mitigated by scaffolding systems, but mortar and other materials can still slip through. Workers have to be aware of the activities taking place above and below them to avoid being hit by falling objects.

SOURCES

NIOSH [2009]. Control of hazardous dust when grinding concrete. Cincinnati, OH: U.S. Department of Health and Human Services, Centers for Disease Control and Prevention, National Institute for Occupational Safety and Health, DHHS (NIOSH) Publication No. 2009–115 [www.cdc.gov/niosh/docs/wp-solutions/2009-115/].

NIOSH [2008]. Water spray control of hazardous dust when breaking concrete with a jackhammer. Cincinnati, OH: U.S. Department of Health and Human Services, Centers for Disease Control and Prevention, National Institute for Occupational Safety and Health, DHHS (NIOSH) Publication No. 2008–127 [www.cdc.gov/niosh/docs/wp-solutions/2008-127/].

Photo courtesy of John Gambatese

WORKPLACE SOLUTIONS

From the National Institute for Occupational Safety and Health

Control of Hazardous Dust When Grinding Concrete

Summary

Construction workers are exposed to hazardous dust when using handheld electric grinders to smooth poured concrete surfaces after forms are stripped. The National Institute for Occupational Safety and Health (NIOSH) found that exposures could be reduced if a local exhaust ventilation (LEV) shroud was attached to the grinder.

Description of Exposure

Breathing dust that contains crystalline silica can lead to the development of silicosis, a deadly lung disease. No effective treatment exists for silicosis, but it can be prevented by controlling workers' exposure to dust containing crystalline silica. Exposure to crystalline silica has also been linked to lung cancer, kidney disease, reduced lung function, and other disorders [NIOSH 2002a].

Workers in the construction industry may breathe dust that contains silica during various tasks including cutting brick and block, tuckpointing masonry, using a jackhammer to break concrete or rock, or grinding concrete. A NIOSH [2001] study found that workers grinding concrete to smooth poured concrete surfaces after forms are stripped were exposed from 35 to 55 times the NIOSH recommended exposure limit (REL) for airborne dust containing crystalline silica. NIOSH evaluated the use of LEV shrouds on handheld concrete grinders to see whether they reduce worker exposure to dust [Echt and Seiber 2002; NIOSH 2002b].

NIOSH Study

The concrete finishers in the NIOSH studies were responsible for smoothing poured concrete walls and columns. The LEV system consisted of a grinder that was equipped with a ventilation shroud, a length of flexible corrugated hose, and a portable electric vacuum cleaner that acted as the fan and dust collector for the ventilation system (Figures 1 and 2). The concrete surfaces were flat and allowed the shroud to make a good seal with the concrete. Four commercially available shrouds were used in the NIOSH study. All grinder/shroud combinations reduced dust exposure by at least 90%.

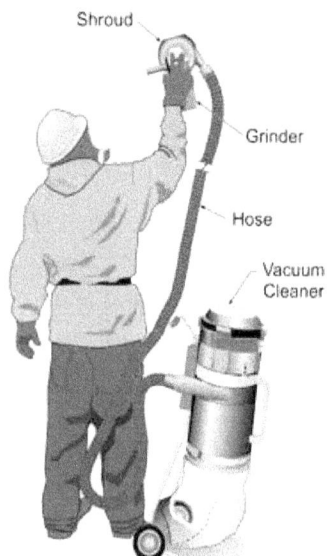

Figure 1. Grinder in use with the control in place.

DEPARTMENT OF HEALTH AND HUMAN SERVICES
Centers for Disease Control and Prevention
National Institute for Occupational Safety and Health

Grinders

The grinders used were rated at either 10,000 or 11,000 rpm: Metabo model W7-115 Quick 10,000-rpm grinder and Metabo model 11025 grinder (Metabowerke GmbH; Nürtingen, Germany); Bosch model 1347A grinder (Robert Bosch GmbH; Stuttgart, Germany); and Milwaukee model 6153–20 grinder (Milwaukee Electric Tool Corp.; Brookfield, WI). The grinders were fitted with 4-inch diameter diamond cup wheels (PW series, Pearl Abrasive Co.; Commerce, CA).

LEV shrouds

Four LEV shrouds were used in the study. The shrouds were selected based on their rugged appearance, how easily they could be mounted on the grinders, and their availability for purchase. The shrouds used were Vacuguard (Pearl Abrasive Co.; Commerce, CA), Dustcontrol (Transmatic Inc; Wilmington, NC); and "full-dust shroud" and "cut (edging) shroud" (Sawtec; Oklahoma City, OK).

Vacuum cleaners and hoses

The grinder/shroud pairs were connected via 1.5-inch (inside diameter) corrugated flexible hose to two types of industrial vacuum cleaners (DC 2700 and DC 3700; Dustcontrol AB, Norsborg, Sweden). The manufacturer reports that the DC 2700 vacuum has a maximum flow capacity of 112 cubic feet per minute (190 cubic meters per hour), and a maximum negative pressure of 84 inches w.g. (21 kPa). The DC 3700 has a maximum flow capacity of 188 cubic feet per minute (320 cubic meters per hour) and a negative pressure of 96.5 inches w.g. (24 kPa).

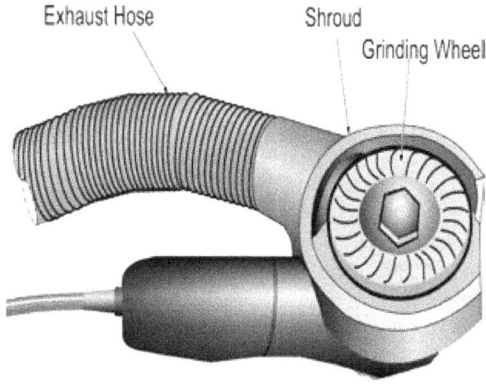

Figure 2. Diagram of grinder showing main parts.

Controls

NIOSH and its partners have developed recommendations to protect workers from exposure to crystalline silica dust during construction activities [NIOSH 1996; Echt and Seiber 2002; NIOSH 2002a; Heitbrink and Collingwood 2005]. Some of the benefits of using the dust control noted in this report include reducing worker exposure to hazardous dust and potentially allowing for use of less protective respiratory protection, reduced cleanup time, and reduced cleanup exposures.

Vacuum cleaners

The choice of a vacuum cleaner depends on the task. It must be carefully selected to include features such as the following:

- Sufficient flow rate to capture the dust and transport it to the vacuum source
- Uses a high-efficiency particulate air (HEPA) filter to reduce the chance of releasing dust containing crystalline silica from the vacuum into the worksite
- Uses pre-filter or cyclone to increase the length of service of the HEPA filter
- Uses filter replacement indicator
- Filters can be cleaned and replaced or full collection bowls or bags can be replaced without exposing the operators to dust

The vacuum cleaner should draw at least 10 amps if it is used as part of a ventilated grinder system, so it can overcome filter loading. Some vacuum cleaners are equipped with a pressure gauge that indicates when the air flow rate is too low to be effective. If the vacuum cleaner does not have a pressure gauge, workers can monitor the air flow by looking at the dust plume. If dust is escaping under the shroud, the dust collected on the pre-filter needs to be dislodged or the vacuum cleaner bags or filters need to be changed.

Hose

A 1.5- or 2-inch diameter hose with a relatively smooth interior and a length of no more than 15 feet should provide adequate air flow. The hose should have as few elbows or turns as possible. A study on tuckpoint grinders [Heitbrink and Collingwood 2005] reported that 2-inch diameter hoses provided better air flow than smaller (e.g., 1.5 inch) diameter hoses. Dust has a greater tendency to settle in larger diameter hoses and should be cleared before and after each use.

Shroud

The shroud can be purchased with the grinder, separately, or as a unit with the vacuum cleaner and hose. The shroud should totally enclose the spaces around the exhaust entry point for the hose. The exhaust shroud should have an entry point for the hose matching the diameter of the hose.

Work practices

- Keep the shroud flat against the surface of the concrete while grinding.
- Shake the hose as needed to loosen the settled dust and prevent the hose from clogging.
- When using the grinder, look to make sure no dust is escaping from the shroud.
- If dust is escaping, turn off the unit and clean or change the filter as recommended by the manufacturer. Sometimes the build-up on the filter can be dislodged by simply moving or shaking the cleaner, or turning the motor off and on a few times. Build-up on the filters slows down the air flow through the system and reduces dust capture.
- Change vacuum cleaner bags before they leak.
- When changing filters, bags, or self-contained collection bowls, use proper disposal practices and use respirators if appropriate.

Since NIOSH last investigated concrete grinders in 2002, several grinder manufacturers have introduced tools with dust controls. Studies since the NIOSH investigation show the effectiveness of LEV controls in reducing respirable dust when using concrete grinders [Croteau et al. 2004; Akbar-Khanzadeh et al. 2007].

Respirators

Workers and employers should be aware of the high risk of dust exposure in poorly ventilated areas (such as in corners or inside buildings). This may result in increased exposure to hazardous dust.

The dust control cited in this report may greatly reduce worker exposure to hazardous dust; however, respirators are still necessary to reduce exposure to crystalline silica below the NIOSH REL of $50\mu g/m^3$. Follow the Occupational Safety and Health Administration (OSHA) Respiratory Protection Standard (29 CFR* 1910.134) (www.osha.gov/SLTC/etools/respiratory/index.html). The provisions of the program include procedures for selection, medical evaluation, fit testing, training, use, and care of respirators.

*Code of Federal Regulations. See CFR in references.

Acknowledgments

The principal contributors to this publication were Alan Echt of the Division of Applied Research and Technology, NIOSH and John J. Whalen under contract with the U.S. Public Health Service, Division of Federal Occupational Health.

References

Akbar-Khanzadeh F, Milz S, Ames A, Susi PP, Bisesi M, Khuder SA, Akbar-Khanzadeh A [2007]. Crystalline silica dust and respirable particulate matter during indoor concrete grinding—wet grinding and ventilated grinding compared with uncontrolled conventional grinding. Occup. Environ. Hyg 4:770–779.

CFR. Code of Federal regulations. Washington, DC: U.S. Government Printing Office. Office of the Federal Register.

Croteau GA, Flanagan ME, Camp JE, Seixas NS [2004]. The efficacy of local exhaust ventilation for controlling dust exposures during concrete surface grinding. Ann. Occup. Hyg 48(6):509–518.

Echt A, Seiber K [2002]. Control of silica exposure from hand tools in construction: grinding concrete. Appl Occup Environ Hyg 17(7):457–461.

Heitbrink WA, Collingwood S [2005]. Protecting tuck-pointing workers from silica dust: draft recommendations for a ventilated grinder. The Center to Protect Workers' Rights, Silver Spring, MD [www.cdc.gov/elcosh/docs/d0600/d000683/d000683.html].

NIOSH [1996]. NIOSH Alert: Request for assistance in preventing silicosis and deaths in construction workers. Cincinnati, OH: U.S. Department of Health and Human Services, Centers for Disease Control and Prevention, DHHS (NIOSH) Publication No. 96–112.

NIOSH [2001]. In-depth survey report of four sites: exposure to silica from hand tools in construction chipping, grinding, and hand demolition at Frank Messer and Sons Construction Company, Lexington and Newport KY; Columbus and Springfield, OH. Cincinnati, OH: U.S. Department of Health and Human Services, Centers for Disease Control and Prevention, National Institute for Occupational Safety and Health, Survey Report No. EPHB 247–15.

NIOSH [2002a]. NIOSH hazard review: Health effects of occupational exposure to respirable crystalline silica. Cincinnati, OH: U.S. Department of Health and Human Services, Centers for Disease Control and Prevention, National Institute for Occupational Safety and Health, DHHS (NIOSH) Publication No. 2002–129.

NIOSH [2002b]. In-depth survey report of control of respirable dust and crystalline silica from grinding concrete at Messer Construction, Newport, Kentucky and Baker Concrete Construction, Dayton, Ohio. Cincinnati, OH: U.S. Department of Health and Human Services, Centers for Disease Control and Prevention, National Institute for Occupational Safety and Health, Survey Report No. EPHB 247–21.

DEPARTMENT OF HEALTH AND HUMAN SERVICES
Centers for Disease Control and Prevention
National Institute for Occupational Safety and Health
4676 Columbia Parkway
Cincinnati, OH 45226–1998

For More Information

The information in this document is based on NIOSH field studies. More information about silica hazards and controls is available on the NIOSH Web site at www.cdc.gov/niosh/topics/silica/default.html.

To receive copies of the NIOSH field study reports that formed the basis of this document or to obtain information about other occupational safety and health topics, contact NIOSH at

Telephone: 1–800–CDC–INFO (1–800–232–4636)
TTY: 1–888–232–6348 ▪ E-mail: cdcinfo@cdc.gov

or visit the NIOSH Web site at www.cdc.gov/niosh

For a monthly update on news at NIOSH, subscribe to *NIOSH eNews* by visiting www.cdc.gov/niosh/eNews.

Mention of any company or product does not constitute endorsement by NIOSH. In addition, citations to Web sites external to NIOSH do not constitute NIOSH endorsement of the sponsoring organizations or their programs or products.

Furthermore, NIOSH is not responsible for the content of these Web sites. All Web sites addressess referenced in this document were accessible as the publication date.

This document is in the public domain and may be freely copied or reprinted. NIOSH encourages all readers of the *Workplace Solutions* to make them available to all interested employers and workers.

As part of the Centers for Disease Control and Prevention, NIOSH is the Federal agency responsible for conducting research and making recommendations to prevent work-related illness and injuries. All *Workplace Solutions* are based on research studies that show how worker exposures to hazardous agents or activities can be significantly reduced.

Control of Hazardous Dust When Grinding Concrete

DHHS (NIOSH) Publication No. 2009–115

SAFER ▪ HEALTHIER ▪ PEOPLE™

April 2009

WORKPLACE SOLUTIONS

From the National Institute for Occupational Safety and Health

Water Spray Control of Hazardous Dust When Breaking Concrete with a Jackhammer

Summary

Construction workers are exposed to hazardous dust when using jackhammers to break concrete pavement. NIOSH found that exposures could be reduced by using a water-spray attachment.

Description of Exposure

Breathing dust that contains crystalline silica can lead to silicosis, a deadly lung disease. Exposure to crystalline silica has also been linked to lung cancer, kidney disease, reduced lung function, and other disorders [NIOSH 2002]. No effective treatment exists for silicosis, but it can be prevented by controlling worker exposure to dust containing crystalline silica.

Workers in the construction industry may breathe dust that contains crystalline silica during many tasks including grinding concrete, cutting brick and block, tuckpointing masonry, or using a jackhammer to break concrete. A study to measure exposures found that jackhammer operators who break concrete were exposed to about 6 times the NIOSH recommended exposure limit (REL) [Valiante et al. 2004]. NIOSH evaluated the use of jackhammers for breaking concrete pavement and examined engineering controls to see whether they reduce worker exposures to dust [Echt et al. 2003].

NIOSH Study

NIOSH studied a water-spray attachment (Figure 1) to suppress dust created during concrete pavement breaking with jackhammers [Echt et al. 2003]. This low-flow, water-spray control reduced dust exposures by 70%–90%.

Figure 1. The water spray attachment, showing the method used to attach the nozzle to the tool.

DEPARTMENT OF HEALTH AND HUMAN SERVICES
Centers for Disease Control and Prevention
National Institute for Occupational Safety and Health

Water-Spray Control

The water-spray attachment was made by a contractor who participated in the NIOSH study (Mt. Hope Rock Products, Inc., Wharton, NJ, a division of Tilcon New York, Inc., West Nyack, NY). There is continuing development of water-spray controls similar to the one used in the NIOSH study. For example, The New Jersey Laborers Health and Safety Fund (NJLHSF) has also developed a simple durable, low-cost water-spray attachment for use on a jackhammer (www.njlaborers.org/index.php3). The NJLHSF version of a water-spray attachment used the control described in this document as a starting point. A detailed description of the NJLHSF water-spray attachment and estimated cost can be found at their Web site (www.njlaborers.com/health/jackhammer.php3).

NIOSH is not aware at this time of off-the-shelf, commercially available retrofit kits or jackhammers that come with built-in water spray units. However, it is relatively simple to build a water spray control for a jackhammer using the diagram in Figure 2 and the parts and instructions below:

- **Water-spray nozzle:** Use a solid-cone, furnace-spray water nozzle with an 80-degree spray angle (Type B, 11.00 GPH, 80°, Delavan Inc. Fuel Metering Products, Bamberg, SC www.delavaninc.com). Mount the nozzle in a bracket welded on the end of the jackhammer. The spray angle (the angle included between the sides of the cone formed by the water discharged by the nozzle) and the spray pattern are two critical design parameters required to match the performance of the tested device. Spray nozzles make several spray patterns such as hollow cone, full cone, and flat spray. This control used a solid cone nozzle.

Figure 2. Diagram of water-spray control used in NIOSH study

- **Water flow rate:** The nozzle used in the NIOSH study delivered about 350 milliliters (11.8 ounces) of water per minute. This flow rate is the third critical design parameter for performance of this control. It is effective in reducing dust and it did not add a lot of water to the work surface or significantly wet workers' clothing or shoes. Higher flow rates may not greatly increase dust control, and lower flow rates may reduce performance.

- **Bracket:** Use a bracket for mounting the water-spray nozzle on the jackhammer. Mounting the nozzle above the end of the jackhammer will prevent the nozzle from striking the pavement.

- **Water-supply lines:** Connect the nozzle by flexible 16 pounds-per-square-inch (psi), 3/8-inch-diameter hydraulic line to a quarter-turn valve mounted near the operator's hand position for turning the water on or off. A 3/8-inch-diameter air hose connects the valve to a 60-gallon water tank (pressurized to 22 psi) mounted on the air-compressor trailer (Figure 3). Control the pressure in the tank with a regulator.

- **Water source:** Use a water tank or a direct connection to a local water supply such as a water main. If a tank is used, water can be supplied to the attachment by pressurizing the tank or pumping water from the tank. If a pressurized tank is used, a compressor is needed to pressurize the tank, a regulator to control the pressure, and a pressure relief valve to guard against the tank bursting. Rust from a steel tank may clog the spray nozzle. However, a plastic water tank can be used with a battery-powered water pump instead of a steel pressurized tank. The larger the tank, the less it will have to be refilled. A 50-gallon tank will easily supply one jackhammer water-spray control used constantly for a full 8-hour shift. A trailer or hand truck may be necessary for moving the tank around the worksite.

Controlling Dust Exposures

The results of the NIOSH study showed that the control devices may reduce exposure to dust for jackhammer operators and other workers near the work area.

Employers and jackhammer operators should take the following steps to reduce worker exposure to hazardous dust:

Site Set-Up

- Develop a site-specific safety and health plan for all job sites where jackhammers are used that considers engineering controls, personal protective equipment, and work practices.

Figure 3. Water tank mounted on compressor trailer

- Minimize the number of workers in the construction area where jackhammers are used.

- Post signs to warn workers about the hazard and to inform them of required protective equipment. Workers should also be cautioned about icing, slips, and falls (particularly if they make a mud hole), and about ground faults for any electrical system in use.

- During jackhammer use, perform air monitoring of respirable crystalline silica exposures to make sure the engineering controls are working and to determine whether workers need respiratory protection.

- Make medical examinations available to all workers exposed to crystalline silica.

Engineering Controls

- Equip jackhammers with dust-reduction control devices such as the water-spray attachment described in this report. When a water-spray attachment cannot be used (for example, on the upper floor inside an occupied building), use other control measures such as a vacuum or other local exhaust ventilation (LEV) device [Echt et al. 2003]. Spraying the work area with a garden hose is not an appropriate replacement for the water-spray control.

- Train workers in the proper use and maintenance of the dust-reduction device. Make sure that the control is working properly and test the water flow rate before and after each shift; a watch with a second hand and a kitchen measuring cup could be used for this task. An 8-ounce cup should fill in about 40 seconds.

Personal Hygiene and Protective Clothing

- Wash hands and face before eating, drinking, or smoking. Do not eat, drink, or use tobacco products in the work area.

- Change into disposable or washable work clothes at the worksite. If possible, shower and change into clean clothes before leaving the worksite. If it is not possible to shower or change into clean clothes, use a vacuum to remove dust from clothes.

- Park cars where they will not be contaminated with silica dust.

- Do not remove dust from the work area by blowing with compressed air or dry sweeping. Also, do not blow dust from clothing or skin with compressed air.

Protective Equipment

- Use hearing and eye protection devices. When water-spray attachments are used with jackhammers, waterproof personal protective equipment may be necessary.

- Use respiratory protection when needed. The controls cited in this report may greatly reduce worker exposure to dust; however, respirators may still be necessary to reduce exposure to crystalline silica below the NIOSH REL of $50 \mu g/m^3$. It may be possible to use less restrictive respirators such as a disposable N–95 filtering facepiece since the amount of hazardous dust is decreased by the controls. The respirators are less cumbersome and cost less than the respirators typically required for jackhammer operators. Employers should follow the Occupational Safety and Health Administration (OSHA) Respiratory Protection Program (29 CFR 1910.134).

Acknowledgments

The principal contributor to this publication was Alan Echt of the Division of Applied Research and Technology, National Institute for Occupational Safety and Health. John J. Whalen, under a NIOSH contract, served as lead writer/editor. This research was conducted in cooperation with the New Jersey Silica Partnership.

References

Echt A, Seiber K, Jones E, Schill D, Lefkowitz, Sugar J, Hoffner K [2003]. Control of respirable dust and crystalline silica from breaking concrete with a jackhammer. Appl Occup Environ Hyg 18:491–495.

NIOSH [2002]. NIOSH hazard review: health effects of occupational exposure to respirable crystalline silica. Cincinnati, OH: U.S. Department of Health and Human Services, Centers for Disease Control and Prevention, National Institute for Occupational Safety and Health, DHHS (NIOSH) Publication No. 2002–129.

Valiante DJ, Schill DP, Rosenman KD, Socie E [2004]. Highway repair: a new silicosis threat. Am J Public Health *94(5)*:876–880.

For More Information

The information in this document is based on NIOSH field studies. More information about silica hazards and controls is available on the NIOSH Web site at www.cdc.gov/niosh/topics/silica/default.html.

To receive copies of the NIOSH field study reports that formed the basis of this document or to obtain information about other occupational safety and health topics, contact NIOSH at

Telephone: 1–800–CDC–INFO (1–800–232–4636)
TTY: 1–888–232–6348 ▪ E-mail: cdcinfo@cdc.gov

or visit the NIOSH Web site at www.cdc.gov/niosh

For a monthly update on news at NIOSH, subscribe to NIOSH *eNews* by visiting www.cdc.gov/niosh/eNews.

Mention of any company or product does not constitute endorsement by NIOSH. In addition, citations to Web sites external to NIOSH do not constitute NIOSH endorsement of the sponsoring organizations or their programs or products.

Furthermore, NIOSH is not responsible for the content of these Web sites.

This document is in the public domain and may be freely copied or reprinted. NIOSH encourages all readers of the *Workplace Solutions* to make them available to all interested employers and workers.

As part of the Centers for Disease Control and Prevention, NIOSH is the Federal agency responsible for conducting research and making recommendations for preventing work-related illnesses and injuries. All *Workplace Solutions* are based on research studies that show how worker exposures to hazardous agents or activities can be significantly reduced.

Water Spray Control of Hazardous Dust When Breaking Concrete with a Jackhammer

DHHS (NIOSH) Publication No. 2008–127

SAFER • HEALTHIER • PEOPLE™   May 2008

DEPARTMENT OF HEALTH AND HUMAN SERVICES
Centers for Disease Control and Prevention
National Institute for Occupational Safety and Health
4676 Columbia Parkway
Cincinnati, OH 45226–1998

More Construction Hazards

- Falls

- Obstructions

- Cave-in during foundation construction

- Lung or skin irritation from exposure to cement or admixtures
  [NIOSH 2008, 2009]

- Jack, cable, or fitting failure during tensioning

Photo courtesy of John Gambatese

Reinforced Concrete

NOTES

Falls and obstructions are common hazards on construction projects. Foundation construction creates holes for potential falls into trenches. Additionally, cave-in from foundation construction, although a well-known hazard, needs to be considered in design. Although there are specific guidelines related to trenching and shoring, they are not always followed.

Lung irritation is a common side-effect of mixing or cutting concrete. Inhaled particles can lead to lung diseases, including silicosis. Skin irritation from exposure to cement or admixtures is possible with prolonged contact to wet concrete. Cements are often strongly basic, which can irritate skin and cause dermatitis. Additionally, as concrete dries on skin, it will draw moisture from the skin.

Concrete post-tensioning operations can be hazardous to construction workers if a jack, cable, or fitting fails during tensioning. The large amount of tension put on each tendon could result in a whiplash effect if the cable snaps.

SOURCES

Photo courtesy of John Gambatese

The NIOSH 2008 and 2009 publications are included in the References.

Construction Industry Statistics [BLS 2011]

Industry	2008 Annual average employment (thousands)	Total recordable cases*	Cases* with days away from work, job transfer, or restriction			Other recordable cases*
			Total	Cases with days away from work	Cases with job transfer or restriction	
Construction	7597.2	4.7	2.5	1.7	0.7	2.2
Poured concrete foundation and structure contractors	235.6	6	3.3	2.3	1	2.8
Structural steel and precast and concrete contractors	105.1	6.4	3.9	2.5	1.4	2.5
Framing contractors	114.5	6.9	4.3	3.1	1.2	2.6
Masonry contractors	231.3	4.6	3.1	2.3	0.8	1.5
Glass and glazing contractors	64.8	7.6	3.4	2.1	1.3	4.1
Roofing contractors	196.2	6.3	3.8	2.7	1.1	2.5
Siding contractors	45.6	5.1	2.5	2.1	0.5	2.6

*Cases per 100 FTE workers

Reinforced Concrete

NOTES

According to the Bureau of Labor Statistics, with regard to nonfatal occupational injuries and illness by industry in 2008, the construction industry as a whole had an incident rate of 4.7 recordable cases per 100 full-time equivalent (FTE) workers. More specifically, for poured concrete foundation and structure contractors, that rate increases to 6.0 recordable cases per 100 FTE workers. This table compares these incident rates with other specialty areas of the construction industry.

SOURCE

BLS [2011]. Injuries, illnesses, and fatalities. Washington, DC: U.S. Department of Labor, Bureau of Labor Statistics [www.bls.gov/iif].

Mitigating Concrete Construction Hazards

REINFORCED CONCRETE DESIGN

Mitigating Concrete Construction Hazards

NOTES

This section provides a number of examples of how PtD can be applied to reinforced concrete design. Concrete structures can vary significantly in size, shape, and configuration. The examples present only a few possible PtD solutions and may not apply to all projects. Thorough review of the design and brainstorming sessions can lead to other PtD elements for a project. This process is facilitated by including persons with construction expertise and knowledge of potential construction site hazards in the review.

PtD Examples

Topic	Slides
Site Activities	48–50
Cranes and Derricks	51
Foundations	52–57
Concrete Floor Surfaces and Elevated Slabs	58–66
Rebar and Post-tensioning Cables	67–69
Formwork	70–71
Concrete Walls, Beams and Girders, and Columns	72–75
Precast Concrete	76–77
Safe Work Procedures	78

NOTES

PtD examples are provided for different concrete structure elements, as shown in the table. In addition to those that apply to the design of the permanent structure, some examples pertain to the design of temporary structures such as formwork. On some projects the engineer may specify construction details and procedures for temporary structures as they impact the performance of the permanent structure. Lastly, examples of how PtD can change construction site work procedures are provided. These examples illustrate how PtD can be extended to design the permanent facility in order to facilitate safer construction procedures.

Site Activities

- Use alternative methods for pouring concrete below or next to overhead power lines

    - Pumping truck

- Consider using onsite batch plant, with inspections performed if required

    - Minimizes transportation hazards

NOTES

These PtD examples relate to getting the concrete to the worksite and placing it on the worksite. Specifications that direct the contractor to minimize transportation requirements and avoid jobsite hazards will lead to safer projects.

 **Site Activities**

Photo courtesy of John Gambatese

Reinforced Concrete

NOTES

This picture shows a pumping truck pouring a foundation. This site has no overhead hazards, but if there were power lines on the site, it would be necessary to avoid them. If at all possible, the power lines should be de-energized during the concrete pumping operation.

SOURCE

Photo courtesy of John Gambatese

Site Activities

- Allow flexibility in concrete mixes. Designate slump and air content ranges and do not preclude adding water at the site.

  - Give the contractor a window of tolerance for less than ideal site conditions such as in poor weather

- Require the constructor to locate and mark existing reinforcing steel prior to cutting into the concrete

  - Preserve the structural integrity of existing reinforced concrete members

NOTES

Unexpected and hidden site conditions can create hazards for the contractor. Flexibility in the design can help the contractor safely conduct construction activities when unexpected conditions arise, while still meeting the design specifications. Respirators should be worn when cutting or mixing concrete.

Cranes and Derricks

- Erection and disassembly must be carefully planned.

- Site layout affects crane maneuverability.

- Show site utilities on plans.

- Comply with OSHA standards.

Photo courtesy of Walter Heckel

The OSHA comprehensive crane standard: www.osha.gov/FedReg_osha_pdf/FED20100809.pdf.
Regulation text: www.osha.gov/cranes-derricks/index.html.

Reinforced Concrete

NOTES

Cranes and derricks are used to lift steel members and equipment into place. Derricks are stationary, for example, when they are built dockside at a port. The six tower cranes shown in this picture were used to transport hoppers filled with concrete to various locations throughout the site. On-site batch plants are often used in urban environments and on large or fast-tracked projects. Cranes are the most complex machines on a construction site. Crane erection and disassembly must be carefully planned.

Where do you place the crane? Ideally, the crane can lift all members from one location without interfering with any other operations. The biggest danger in site layout is overhead power lines. Although it is the contractor's responsibility to deal with power lines, the designer can help by including the power line locations on the plans.

Another problem, overturning, is often the result of moments created by the load. Cherry pickers are particularly susceptible. Cranes operate within a range defined by the mass of the crane, the length of the boom, and the mass of the load. Operators may be tempted to extend the boom a few more feet to pick up a load, when it would be safer to move the crane closer. As the load is lifted, the crane tips.

Another hazard is boom collapse. In this instance, the lift exceeds the design limits of the boom. There is always the possibility that the operator will lose control of the load, especially when it is windy. A swinging load may impact adjacent structures or touch a power line. In several instances, the crane operator has died when the load swung back into the cab.

SOURCES

OSHA [2010]. Comprehensive crane standard [www.osha.gov/FedReg_osha_pdf/FED20100809.pdf].

OSHA Regulations for cranes and derricks [www.osha.gov/cranes-derricks/index.html].

Photo courtesy of Walter Heckel

Foundations

- Do not use driven piles in deep excavations in areas of loose or backfilled soil

    - Prevent cave-ins

- Avoid designing piles at angles flatter than 4:12 (horizontal: vertical)

- When developing a plot plan, group footings in a way that permits proper drainage of mass excavations

    - Avoid water build-up on site

NOTES

Many construction site accidents associated with foundation construction occur because of cave-in. Foundation designs can mitigate cave-in hazards if minimal soil vibration occurs and water is able to drain freely. Battered piles can also cause safety hazards because of the horizontal nature of the implied forces.

 **Foundations**

- Use 4" × 4" mat mesh or welded wire fabric (WWF) on top of more widely spaced top rebar
  - Provides walking surface
- Review clearances between forms, anchor bolts, sleeves, and rebar at congested pier locations
  - Ensure sufficient room for equipment
- Standardize anchor bolts to several different diameters, types, and lengths
  - Prevent confusion about placement

Reinforced Concrete

NOTES

Providing a worksite that is free of tripping and walking hazards can be difficult. During the construction of concrete foundations, the workers must walk across the exposed rebar prior to pouring of the concrete. To eliminate tripping hazards, rebar can be designed to provide an easier walking surface. Allowing room for workers to maneuver is another important design practice. Consider how the workers are going to install the rebar, anchors, etc. Allow room for equipment used to place and consolidate the concrete. Making a work area easily accessible goes a long way in providing a safe worksite. Standardizing all elements of the design is good practice. Using standard features eliminates confusion and the need for rework.

 **Foundations**

Photo courtesy of John Gambatese

Reinforced Concrete

NOTES

This worker has to balance on the rebar while vibrating the concrete for the foundation. Design the rebar with a maximum 4" spacing to allow him to walk without worrying about falling through the spaces.

SOURCE

Photo courtesy of John Gambatese

 **Foundations**

- Design placement directly against earth, instead of forming, where conditions permit
  - Prevent formwork blowouts
- Design small foundations and slabs-on-grade without haunches
  - Irregular, small excavated areas can be tripping hazards
- Eliminate offsets, tapered sections, and other complicated shapes
  - Cave-in hazards

NOTES

Simplifying the formwork and foundation design is another means of facilitating safe construction. Irregular shapes can lead to complicated formwork and unexpected site conditions while the foundation is being constructed. As much as possible, the engineer should use simple shapes that do not have offsets, tapered sections, or other varied dimensions.

Foundations

- Design-in adequate embedment in concrete foundations, piers, and walls
    - Allows easy attachment of platforms, stairs, light fixtures, etc.
- Provide railing or grating on top of sumps
    - Prevents falls into the sump pit

Photo courtesy of Thinkstock

Reinforced Concrete

NOTES

Consideration should be given to the design features that will be added to the concrete after it is cast. These may include steel stairs, cable trays, light fixtures, and other mechanical and electrical pieces of equipment. Facilitating the installation of the added features will go a long way in helping the contractor during their installation.

Fall hazards are especially important in construction, because falls are the leading cause of injuries and fatalities. Protecting exposed sump pits as part of the permanent design can eliminate potential exposure to this hazard during construction.

SOURCE

Photo courtesy of Thinkstock

Foundations

- Standardize foundation sizes for pumps, pipe racks, structures, and miscellaneous supports
    - Standard, regular work environment helps workers
- Dimension concrete foundations and structures to maximize use of commercial form sizes
    - Custom forms may be under-designed or difficult to install

NOTES

Create a design that facilitates familiarity with the site. Standardize the worksite to reduce hazards. This ensures that workers will not be surprised by unexpected conditions as they work. Avoid custom work; specify standard sizes and shapes.

 **Concrete Floor Surfaces**

- Keep steps, curbs, blockouts, slab depressions, and other similar floor features away from window openings, exterior edges, and floor openings

- Design the covers over sumps, outlet boxes, drains, etc., to be flush with the finished floor

- Provide a non-slip walking surface on walkways and platforms that are adjacent to open water or exposed to the weather

Reinforced Concrete

NOTES

Slipping and tripping hazards are especially troublesome during construction of floor surfaces. Keep the floor surface free of protruding elements, holes, and standing water.

 **Concrete Floor Surfaces**

Photo courtesy of John Gambatese

NOTES

This picture shows a tripping hazard (exposed rebar) directly adjacent to a fall hazard (change in walking surface elevation).

SOURCE

Photo courtesy of John Gambatese

Concrete Floor Surfaces

- For access doors through floors, use doors which immediately provide guarded entry around the whole perimeter when the door is opened

- Locate floor openings away from passageways, work areas, and the structure perimeter

- Eliminate tripping hazards (changes in elevation, curbs, etc.) around floor openings

NOTES

Floor openings are significant hazards during construction. Consider the location of the openings, their size, and their shape. Where possible, locate the openings away from places where construction workers are likely to walk. Make them as small as possible to prevent falling through the openings, and cluster them together so that they can be easily guarded.

Concrete Floor Surfaces

- Specify broom finish (non-slip walking surfaces) on floors adjacent to open water or exposed to the weather.

- For slabs-on-grade, specify the compaction requirements of the backfill around foundations. Schedule backfilling completion as soon as possible.

NOTES

A broom finish on concrete floors provides a nonslip walking surface. Broom finishes on ramps in parking garages are also a good idea. Backfill provides additional support for slabs-on-grade during the curing process and may prevent or lessen cracking.

Elevated Slabs

- Provide drainage for all floor areas, especially around elevated equipment pads.

- Prohibit the manual placement of metal decking or forms, especially on elevated structures, if wind speeds exceed 25 mph.

- Provide permanent guardrails around floor openings.

NOTES

Concrete construction work is exposed to the environment. Environmental conditions can create safety hazards during concrete construction operations. Consider the weather.

A contractor commonly provides temporary guardrails around openings during construction. However, temporary guardrails may not be installed adequately to meet standards. Specify the installation of permanent guardrails around floor openings to eliminate this problem.

 **Elevated Slabs**

- Note on the contract drawings the existing and new floor design loads
  - Help the constructor in determining material stockpile locations and heavy equipment maneuverability

Photo courtesy of Thinkstock

- For elevated floors, use permanent metal-formed decking with concrete fill to eliminate temporary formwork

Reinforced Concrete

NOTES

Work on elevated floors requires stockpiling materials on the floors prior to their installation. The contractor must know the floor design loads to determine the amount of materials that can be safely stockpiled. Metal formed decking with concrete fill is a fast and efficient construction application. The metal decking can be easily installed and stays in place. Hazards related to temporary formwork and removing formwork are eliminated with the use of metal decking. Samples of concrete from the pour must still be tested to verify strength!

SOURCE

Photo courtesy of Thinkstock

 **Elevated Slabs**

- When showing pipe sleeves on drawings, consider whether the sleeves will be installed before or after the concrete is placed
  - Prevent unnecessary rework at elevated locations after the concrete is in place

- When specifying a top-of-concrete elevation, consider the combined steel and concrete tolerance (including deflection)
  - This may influence the beam size, composite design of floor, and F_f and F_l numbers for floor flatness and levelness.

NOTES

 **Elevated Slabs**

Photo courtesy of John Gambatese

Reinforced Concrete

NOTES

This picture shows typical slab formwork and shoring. The flat slab, without depressions, offsets, etc., provides a smooth and consistent walking surface across the formwork during construction.

SOURCE

Photo courtesy of John Gambatese

 **Elevated Slabs**

- Design concrete members to be of similar size and regularly spaced to facilitate the use, and re-use, of pre-fabricated forms.

- Minimize the number of details to reduce costs and construction errors.

- Consider using bent steel-form plate around the edges of concrete slabs at large openings and around the perimeter.

  - Keep rebar installers away from exposed edges.

- Specify composite steel-form deck.

  - Eliminate formwork and minimize rebar in elevated slabs.

NOTES

If possible, designs should eliminate the need for workers to be close to exposed edges or in elevated locations.

Post-tensioning Cables

- Align or locate post-tensioning cables such that if failure of a jack, cable, or fitting occurs during tensioning, the cable is not directed towards an active work area.

NOTES

When post-tensioning cables and fittings that are highly tensioned fail, they have been known to shoot out of the structure and cause great damage. If possible, design the structure to eliminate potential damage if a failure occurs.

Much work has been done to identify how to design concrete structures for efficient form design. This not only eliminates unnecessary field work but also saves on construction costs and improves the quality of the work. Utilizing prefabricated forms eliminates the need to install custom forms on-site and therefore reduces construction hazards.

Rebar

- Show splice location and splice lengths on the drawings

- Standardize use of a few sizes of rebar such as #5, #7, and #10
  - Between bars that are of similar size
  - Two smaller sizes can substitute for one larger size if field conditions warrant

- Where practical, show vertical wall and pier dowels extending to 6' height instead of using vertical bars spliced to the dowels

NOTES

Consideration of rebar size selection and detailing can ensure easy, safe installation.

 **Rebar**

- Use one grade of rebar throughout the whole job

- Prefabricate column and wall cages when feasible

- Utilize welded wire fabric (WWF) (flat sheets) for area paving reinforcement

- Specify carbon microfibers where design allows

Photo courtesy of Thinkstock

Reinforced Concrete

NOTES

The slide contains additional rebar detailing suggestions to enhance worker safety. When more than one grade of rebar is used, mistakes are likely to happen. For large projects, it may be economically feasible to order prefabricated rebar cages for columns or wall segments. Welded wire fabric (WWF) may be used for large areas of pavement. Additional inspections of structural connections are required to confirm that the WWF is sufficiently anchored with column steel. Carbon fibers have the potential to replace reinforcing steel in some concrete applications.

SOURCE

Photo courtesy of Thinkstock

Formwork

- It is customary to prohibit forming work by hand if wind speed exceeds 25 mph

- Limit the lift height of concrete pours to minimize the load on formwork and the risk of collapse of fresh concrete during pouring operations

Photo courtesy of John Gambatese

Reinforced Concrete

NOTES

The placement of large formwork sections can be hazardous, especially when it is windy. Specifications can guide the contractor toward safe construction methods. In this picture, a tower crane is moving a wall formwork section. Design walls to encourage the reuse of forms, which saves time and reduces cost.

SOURCE

Photo courtesy of John Gambatese

Formwork

- For complicated and large formwork designs, specify that formwork calculations and drawings must be reviewed and stamped by a licensed engineer

- Specify the minimum compressive strength for removal of elevated forms if different than the design compressive strength of the concrete

    - Prevents collapse of the structure due to early removal of the forms

NOTES

Requiring professional engineering review and approval of formwork designs adds an additional level of confidence that the formwork is designed adequately. Clearly communicating to the contractor the minimum concrete compressive strength that must be attained prior to removing the formwork can help prevent collapse of the concrete.

 **Concrete Walls**

- Use one or more curtains of WWF for reinforced concrete walls and columns
  - Allows placement of large sections rather than many small pieces

Photo courtesy of John Gambatese

Reinforced Concrete

NOTES

Building wall rebar up above the floor surface requires constructors to work at elevation. Use of prefabricated rebar sections and panels of WWF eliminates some of the work at elevation.

SOURCE

Photo courtesy of John Gambatese

Concrete Beams and Girders

- Design members of consistent size and shape
    - Standardize the work environment

- Specify a minimum beam width of 6 inches
    - Provides a wide walking surface

- Minimize the use of cantilevers, which can be hard to form and finish.

- Design pre-fabricated members to be of one size and shape, or make them easily distinguishable to avoid incorrect placement.

NOTES

Standardization aids in simplifying construction of concrete elements and in creating a safe worksite. Simple structural elements are easier to construct and require less work to form and finish.

 **Concrete Beams and Girders**

- Design concrete members to be of similar size and regularly spaced

    - Facilitates the use, and reuse, of prefabricated forms

- Consider using shotcrete instead of poured concrete

    - Does not require a form on one side of the member

- Design member depths to allow adequate head room clearance around stairs, platforms, valves, and all areas of egress.

American Concrete Institute
www.shotcrete.org

NOTES

Shotcrete is an all-inclusive term for sprayed concrete or mortar in a dry-mix or wet-mix process. The use of the term *shotcrete* first occurred in *Railroad Age* magazine more than 50 years ago, in place of the then-proprietary word *Gunite*, and has been used by the American Concrete Institute since at least 1967 to describe all sprayed concrete or mortar.

SOURCE

American Concrete Institute [www.shotcrete.org].

Concrete Columns

- Design columns with holes (sleeves) or embedded attachment points for guardrails and lifelines

- Specify long rebar lengths to minimize rebar splices

Photo courtesy of Thinkstock

Reinforced Concrete

NOTES

Design columns with holes in vertical members at approximately 21" and 42" above the floor to install guardrails or lifelines. In those portions of the project where steel decking is to be installed, another hole at 7 ft above the deck allows the use of horizontal lifelines for fall arrest during leading-edge decking. In communities subject to seismic activity, detailing requirements must conform to local codes to ensure adequate seismic performance.

SOURCE

Photo courtesy of Thinkstock

Precast Concrete

- Maximize the use of pre-cast manholes, pull boxes, and other miscellaneous concrete items.

- For precast concrete members, provide inserts or other devices to attach lines or lanyards for fall protection.

Reinforced Concrete

NOTES

Utilizing precast concrete members reduces onsite work at elevation. Precast elements are more consistent, which enables easier installation. Embedded items can be installed easier and located in exactly the right locations, eliminating rework in the field.

Precast Concrete

Photo courtesy of John Gambatese

Reinforced Concrete

NOTES

Use of precast vaults (pictured above) and manholes, for example, is cost-effective and reduces the need for creative forms and extra core drilling. Eliminating the need for formwork and core drilling enhances worker safety on the jobsite.

SOURCE

Photo courtesy of John Gambatese

Safe Work Procedures

- Specify that the device must be embedded in concrete members when testing strength before form removal.

- Design scaffolding tie-off points into exterior walls of buildings for construction purposes.

- Design special attachments or holes in structural members at elevated work areas to provide permanent, stable connections for supports, lifelines, guardrails, scaffolding, or lanyards.

Reinforced Concrete

NOTES

The concrete design can affect the safe installation and use of construction systems such as scaffolding and guardrails. Designing the embedment for anchorage points enables the contractor to tie-off lanyards, support scaffolding, and install other features immediately after the concrete is cured. This ability prevents the need to have workers install the anchorage points after the fact.

Case Study

REINFORCED CONCRETE DESIGN
Construction Case Study

NOTES

This case study example is provided to illustrate how the design of a concrete structure impacted the work of the contractor and ultimately contributed to injuries and fatalities. The case study comes from a 2003 OSHA accident investigation.

Construction Case Study

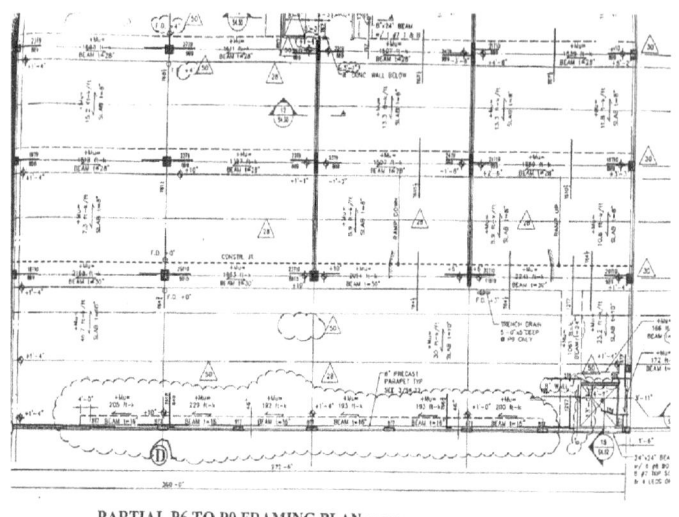

PARTIAL P6 TO P9 FRAMING PLAN (CONTRACT DRAWING S-1.36, REV. 50)
(NTS)

Drawing courtesy of OSHA

Reinforced Concrete

NOTES

This case study is based on an OSHA accident report for an incident that happened on October 30, 2003. A company was in the process of expanding its facilities. The project included a 31-story hotel and a 10-story garage, 8 stories of which were for parking. The framing plan above shows the beam, column, and slab intersection at a typical elevated floor level along one exterior edge of the garage as it was originally designed by the structural engineer. Individual bars were used in the slab and detailed as shown to extend into the column. The beam rebar also extended through the column. The beam shape is such that its depth is greater than its width.

SOURCE

Drawing courtesy of OSHA

Comparison of Design and As-Built

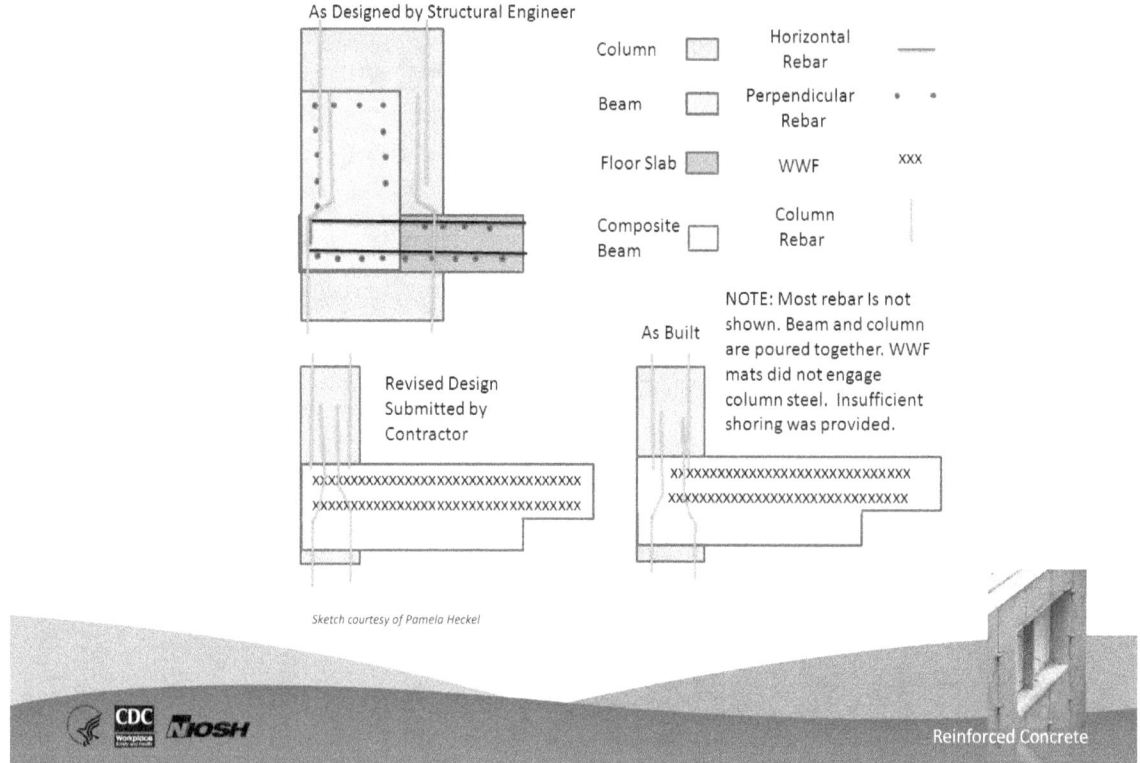

Sketch courtesy of Pamela Heckel

Reinforced Concrete

NOTES

Construction of the lower floors followed the original design. As illustrated, the column was deeper than the beam was wide. This difference created inefficiencies for the contractor in designing and installing the formwork to build the structure. For the upper floors, the contractor submitted a revised design. As shown, the column is much thinner. The separate beam and slab were replaced with a composite beam and floor system, cantilevered from the column. The metal decking in the composite floor system was used to support the weight of the wet concrete during construction and carry tensile stresses in the bottom of the slab. Unfortunately, the stress distribution in a cantilevered beam is reversed, with compression on the bottom and tensile stress on the top. The beam rebar was modified to be a mat of WWF rather than individual bars.

With the original design, the workers threaded the individual bars from the floor slab into the columns, engaging the slab steel with the column steel as required by the American Concrete Institute (ACI) code. With the new design, however, it was difficult to anchor the WWF mats into the column steel. The rebar mats were harder to handle and the columns were thinner. There were indications that the composite beam and floor systems failed to attain strength before additional upper floors were poured. In the weeks before the incident, workers noticed an ominous deflection in shore posts and attempted to alert management, according to newspaper reports.

SOURCE

Sketch courtesy of Pamela Heckel

Case Study—Construction Failure

Photo courtesy of OSHA

Reinforced Concrete

NOTES

Due to difficulties associated with constructing the alternate design, on some of the upper floors, the rebar mats were short of the column steel. As a result, there was no continuity from the slab steel to the column steel, and tension forces in the top steel were not able to be resisted by the column. The upper floor levels collapsed during construction, killing four workers and injuring 20 others.

What aspects of PtD were not utilized in the original drawings?

There was no trade contractor input during the design phase. A constructability review of the original drawings might have identified the costs associated with the beam design. Requiring the column and beam to be the same width would have made the formwork less expensive to build and easier to construct and would have facilitated the construction process.

What aspects of PtD were not utilized in the revised drawings?

The structural engineer of record, a health and safety professional, the designer, and an experienced trade representative could have conducted a constructability review of the final design, which showed a composite beam with WWF. A contractor experienced with WWF might have recommended a different beam-to-column connection.

How could the implementation of PtD have helped to avoid the collapse?

The PtD process shown on slide 15 involves health and safety professionals and trade representatives in the design phase. A health and safety review of the modified design drawings may have identified the potential for "hand traps" and other constructability issues, given the tight fit between WWF and column steel.

What would alternate details look like if following the principles of PtD?

In this instance, the WWF used as horizontal reinforcing steel for the composite beam made an insufficient connection with the column steel. An alternate design as a result of a constructability audit may have resolved that problem.

What is another advantage of PtD?

Lives are saved and accidents are averted when there is a culture of safety awareness. In this instance, warnings from workers about the deflection of slabs and shores were ignored.

SOURCES

Lipton E [2004]. Design changes preceded collapse of casino garage. New York Times, April 25 [www.nytimes.com/2004/04/25/nyregion/25xcollapse.html].

Photo courtesy of OSHA

Summary

 **Recap**

- Prevention through Design (PtD) is an emerging design process for saving lives, time, and money.

- PtD is the smart thing to do and the right thing to do.

- Although site safety is the contractor's responsibility, the designer has an ethical duty to create drawings with good constructability.

- There are tools and examples available to facilitate PtD in reinforced concrete design.

Reinforced Concrete

NOTES

Prevention through Design saves lives, time, and money. PtD is the ethical thing to do. Good constructability is the designer's responsibility.

Help make the workplace safer...

Include *Prevention through Design* **concepts in your projects.**

For more information, please contact the National Institute for Occupational Safety and Health (NIOSH) at

Telephone: (513) 533–8302
E-mail: preventionthroughdesign@cdc.gov

Visit these NIOSH Prevention through Design Web sites:

www.cdc.gov/niosh/topics/PtD

www.cdc.gov/niosh/programs/PtDesign

Reinforced Concrete

NOTES

This presentation was intended to provide examples of construction hazards and risks that could be positively or negatively affected by design decisions. It is certainly not comprehensive in any way. All members of the construction project team (owner, designers, contractors, and safety professionals) must attempt to learn more about construction site safety early in the built environment's life cycle. The earlier more is learned, the more effective and safer the process can be. Each party has a role to play. The United Kingdom and Australia have promulgated designers' roles and responsibilities for safe construction design. Those designers are still learning how to identify and manage risks and how they can provide safer and healthier designs. We encourage the infusion of construction and safety knowledge into the design team and design reviews. Organizations and individuals seeking to positively impact construction workers' safety and health through design will need first an open mind and second a holistic view of what factors influence workers' actions and inactions. Are there questions?

SOURCE

NIOSH Prevention through Design Program Web sites:

www.cdc.gov/niosh/topics/PtD/

www.cdc.gov/niosh/programs/PtDesign/

References

American Institute of Industrial Hygienists [AIHA] [2008]. Strategy to demonstrate the value of industrial hygiene [www.aiha.org/votp_NEW/pdf/votp_exec_summary.pdf].

ANSI/AIHA [2005]. American national standard for occupational health and safety management systems. New York: American National Standards Institute, Inc. ANSI/AIHA Z10-2005.

Behm M [2005]. Linking construction fatalities to the design for construction safety concept. Safety Sci 43:589–611.

BLS [2003–2009]. Census of Fatal Occupational Injuries. Washington, DC: U.S. Department of Labor, Bureau of Labor Statistics [www.bls.gov/iif/oshcfoi1.htm].

BLS [2003–2009]. Current Population Survey. Washington, DC: U.S. Department of Labor, Bureau of Labor Statistics [www.bls.gov/cps/home.htm].

BLS [2006]. Injuries, illnesses, and fatalities in construction, 2004. By Meyer SW, Pegula SM. Washington, DC: U.S. Department of Labor, Bureau of Labor Statistics, Office of Safety, Health, and Working Conditions [www.bls.gov/opub/cwc/sh20060519ar01p1.htm].

BLS [2011a]. Census of Fatal Occupational Injuries. Washington, DC: U.S. Department of Labor, Bureau of Labor Statistics [www.bls.gov/news.release/cfoi.t02.htm].

BLS [2011b]. Injuries, illnesses, and fatalities (IIF). Washington, DC: U.S. Department of Labor, Bureau of Labor Statistics [www.bls.gov/iif/home.htm].

CFR. Code of Federal Regulations. Washington, DC: US Government Printing Office, Office of the Federal Register.

CPWR [2008]. The construction chart book. 4th ed. Silver Spring, MD: Center for Construction Research and Training.

Driscoll TR, Harrison JE, Bradley C, Newson RS [2008]. The role of design issues in work-related fatal injury in Australia. J Safety Res 39(2):209–214.

European Foundation for the Improvement of Living and Working Conditions [1991]. From drawing board to building site (EF/88/17/FR). Dublin: European Foundation for the Improvement of Living and Working Conditions.

Gambatese JA, Hinze J, Haas CT [1997]. Tool to design for construction worker safety. J Arch Eng 3(1):2–41.

Hecker S, Gambatese J, Weinstein M [2005]. Designing for worker safety: moving the construction safety process upstream. Prof Saf 50(9):32–44.

Hinze J, Wiegand F [1992]. Role of designers in construction worker safety. Journal of Construction Engineering and Management 118(4):677–684.

Lipton E [2004]. Design changes preceded collapse of casino garage. New York Times, April 25 [www.nytimes.com/2004/04/25/nyregion/25xcollapse.html].

Lipscomb HJ, Glazner JE, Bondy J, Guarini K, Lezotte D [2006]. Injuries from slips and trips in construction. Appl Ergonomics *37*(3):267–274.

Main BW, Ward AC [1992]. What do engineers really know and do about safety? Implications for education, training, and practice. Mechanical Engineering *114*(8):44–51.

New York State Department of Health [2007]. A plumber dies after the collapse of a trench wall. Case report 07NY033 [www.cdc.gov/niosh/face/pdfs/07NY033.pdf].

NIOSH [2008]. Water spray control of hazardous dust when breaking concrete with a hammer. Cincinnati, OH: U.S. Department of Health and Human Services, Centers for Disease Control and Prevention, National Institute for Occupational Safety and Health, DHHS (NIOSH) Publication No. 2008–127 [www.cdc.gov/niosh/docs/wp-solutions/2008-127/].

NIOSH [2009]. Control of hazardous dust when grinding concrete. Cincinnati, OH: U.S. Department of Health and Human Services, Centers for Disease Control and Prevention, National Institute for Occupational Safety and Health, DHHS (NIOSH) Publication No. 2009–115 [www.cdc.gov/niosh/docs/wp-solutions/2009-115/].

NIOSH [2010]. Reducing work-related musculoskeletal disorders among rodbusters. Cincinnati, OH: U.S. Department of Health and Human Services, Centers for Disease Control and Prevention, National Institute for Occupational Safety and Health, DHHS (NIOSH) Publication No. 2010–103 [www.cdc.gov/niosh/docs/wp-solutions/2010-103/].

NIOSH Fatality Assessment and Control Evaluation (FACE) Program [1983]. Fatal incident summary report: scaffold collapse involving a painter. FACE 8306 [www.cdc.gov/niosh/face/Inhouse/full8306.html].

NOHSC [2001]. CHAIR safety in design tool. New South Wales, Australia: National Occupational Health & Safety Commission.

OSHA [2001]. Standard number 1926.760: fall protection. Washington, DC: U.S. Department of Labor, Occupational Safety and Health Administration.

OSHA [ND]. Fatal facts accident reports index [foreman electrocuted]. Accident summary no. 17 [www.setonresourcecenter.com/MSDS_Hazcom/FatalFacts/index.htm].

OSHA [ND]. Fatal facts accident reports index [laborer struck by falling wall]. Accident summary no. 59 [www.setonresourcecenter.com/MSDS_Hazcom/FatalFacts/index.htm].

Szymberski R [1997]. Construction project planning. TAPPI J *80*(11):69–74.

Toole TM [2005]. Increasing engineers' role in construction safety: opportunities and barriers. Journal of Professional Issues in Engineering Education and Practice *131*(3):199–207.

USC. United States Code. Washington, DC: US Government Printing Office.

Other Sources

American Concrete Institute: www.concrete.org/general/home.asp.

ASSE [2004]. ANSI A10 National standards for construction and demolition operations: A10.9 concrete and masonry construction [www.asse.org/cartpage.php?link=standards].

American Society of Civil Engineers [ASCE]: www.asce.org/.

Gambatese JA [1996]. Addressing construction worker safety in the project design. PhD Dissertation, Department of Civil Engineering, University of Washington, Seattle, WA.

NIOSH Fatality Assessment and Control Evaluation Program: www.cdc.gov/niosh/face/.

National Society of Professional Engineers [NSPE]: www.nspe.org/ethics.

NIOSH Prevention through Design Web sites:
www.cdc.gov/niosh/topics/PtD/
www.cdc.gov/niosh/programs/PtDesign/

OSHA Fatal Facts: www.osha.gov/pls/publications/publication.AthruZ?pType=AthruZ

OSHA home page: www.osha.gov/pls/oshaweb/owastand.display_standard_group?p_toc_level=1&p_part_number=1926.

OSHA PPE publications:
www.osha.gov/Publications/osha3151.html
www.osha.gov/OshDoc/data_General_Facts/ppe-factsheet.pdf
www.osha.gov/OshDoc/data_Hurricane_Facts/construction_ppe.pdf

Test Questions

1. What is the goal of Prevention through Design?

2. Give two examples of industries that have incorporated PtD into the corporate culture.

3. Name one practical benefit of PtD.

4. Give one ethical reason for PtD.

5. Give an example of a hazard associated with an urban construction site.

6. What conditions might cause the sides of an excavation to cave in?

7. List three kinds of personal protective equipment (PPE).

8. Give three reasons why PPE is considered the solution of last resort.

9. How is PtD different from engineering controls?

10. Define constructability.

11. Name the players who must communicate during the design phase.

12. When in the design process is the time to consider safety?

13. Why should you visit the OSHA Web site?

14. Name three construction hazards.

15. Where can you find tools to help you create safer designs?

Answers

1. The goal of PtD is to anticipate and eliminate hazards and risks at the design phase of a project/process and to make workplaces safer for workers.

2. Construction companies, computer and communications corporations, design-build contractors, electrical power providers, engineering consulting firms, oil and gas industries, water utilities

3. Accidents on the job hurt employee morale, delay project completion, and cost money.

4. Preventable accidents should be prevented! Accidents ruin lives.

5. Examples include overhead power lines, existing infrastructure (gas, electric, and sewer), pedestrians, and traffic flow.

6. A trenching accident may be caused by spring thaw, lack of shoring, cracked forms, recent precipitation, type of soil, or placement of heavy equipment.

7. Personal Protective Equipment, or PPE, includes items worn as a last line of defense against injury. OSHA-required PPE can include hardhats, steel-toed boots, safety glasses or safety goggles, gloves, earmuffs, full body suits, respiratory aids, face shields, and fall harnesses.

8. PPE is a solution of last resort because it
    a. requires the worker to wear it,
    b. may not fit because of limited size availability, and
    c. does not eliminate the hazard.

9. Engineering controls isolate the process or contain the hazard. PtD removes or reduces the hazard.

10. The term *constructability* implies an evaluation of a particular design in terms of cost, safety, duration, and quality. Can the design be built at a reasonable cost, within a reasonable amount of time, and result in an acceptable level of quality?

11. The entire design team must communicate, including the architect, structural engineer, civil engineer, HVAC engineer, trade representatives, and site planner.

12. Throughout!

13. OSHA regulations are updated annually. The Web site contains a summary of the latest hazard investigations. It also contains information about occupational diseases.

14. Hazards include falls, tripping hazards, falling objects, loud noises, and musculoskeletal injuries.

15. Agencies such as OSHA, NIOSH, and CHAIR can provide tools to help you create safer designs.